DEVELOPMENTAL DISABILITIES - MOLECULES INVOLVED, DIAGNOSIS, AND CLINICAL CARE

Edited by **Ahmad Salehi**

Developmental Disabilities - Molecules Involved, Diagnosis, and Clinical Care
http://dx.doi.org/10.5772/56808
Edited by Ahmad Salehi

Contributors

Peining Li, Fang Xu, Danilo Moretti-Ferreira, Stanislava Mandzáková, Nathalie G. Berube, Adrienne Elbert, Uner Tan, Sibel Karaca, Meliha Tan

Published by InTech

Janeza Trdine 9, 51000 Rijeka, Croatia

Publishing Process Manager Iva Simcic
Technical Editor InTech DTP team
Cover InTech Design team

Additional hard copies can be obtained from orders@intechopen.com

Developmental Disabilities - Molecules Involved, Diagnosis, and Clinical Care, Edited by Ahmad Salehi
p. cm.
ISBN 978-953-51-1177-1

For EU product safety concerns:
IN TECH d.o.o., Prolaz Marije Krucifikse Kozulić 3, 51000 Rijeka, Croatia,
info@intechopen.com or visit our website at intechopen.com.

Meet the editor

Ahmad Salehi, M.D., Ph.D. is a Clinical Associate Professor at the Department of Psychiatry and Behavioral Sciences, Stanford Medical School and at the VA Palo Health Care System in California. His main interest in the field of developmental disabilities is Down syndrome, using mouse models and human postmortem material. Due to his innovative use of mouse models to identify genes responsible for degeneration of different brain regions in Down syndrome, in 2010, he received the World Technology Award in the field of biotechnology from the World Technology Network in Manhattan, New York. Currently, his group is working on developing new therapeutic strategies for cognitive dysfunction in people with Down syndrome.

Contents

Preface

Millions of children and adults around the world are affected by developmental disabilities. This has affected the cognitive, motor, and emotional aspects of their lives. Although there has been an enormous progress in our understanding on the pathophysiological aspects of neurodevelopmental disorders, many aspects of these disorders remain unknown. For instance, there are 6 millions people affected by Down syndrome around the word. While advance in clinical care and surgical techniques have led to a drastic increase in life expectancy of people with Down syndrome, no effective treatments are available to restore cognitive function in people affected. For this reason there is an urgent need to better understand mechanisms behind cognitive disability and develop new therapies for Down syndrome and similar neurodevelopmental disorders.

The propose of the current book is to provide examples of methods of understanding the molecular mechanisms, clinical aspects, diagnosis, and management of behavioral abnormalities in people with intellectual disability.

The first chapter by Fang Xu and Peining Li discusses the role of cytogenomic abnormalities in people with intellectual disability. Following extensive review of methodology used to perform cytogenetic and genomic analyses, they have discussed in vitro cellular phenotyping and in vivo animal modeling for neurodevelopmental abnormalities.

In the second chapter Adrienne Elbert and Nathalie Berube talk about the role of chromatin structure in people with intellectual disability. The chapter starts with the basic concept of chromatin organization and provides a few examples in which altered chromatin structure plays a significant role in the occurrence of the phenotype.

Chapters 3 and 4 provide examples of rare forms of cognitive disability in adults. These abnormalities are a combination of a variety of motor and cognitive dysfunction. In Chapter 3, Danilo Moretti-Ferreira describes magenis syndrome i.e. a rare form of intellectual disability with extensive overlap with other forms of intellectual disability particularly Down syndrome. This will be followed by a chapter by Karaca, Tan, and Tan describing a rare form of intellectual disability with very characteristic motor problems.

The last chapter (Chapter 5) is dedicated to studying the effects of sexual maturation in people with sever and moderate intellectual disability. In this extensive chapter, Stanislava Mandzáková walks us through behavioral aspects of sexual maturation in people with developmental abnormalities and tests the effects of different factors on sexual behaviors in people affected and their care givers.

Advance in automated full sequencing systems; enabling us to perform genotyping in a large number of individuals, sophisticated imaging systems; providing an ability to accurately study minor alterations in brain circuits, and further availability of high throughput assay systems; to test the therapeutic and toxic effects of an incredibly large number of chemicals in vitro, have further accelerated the process of understanding the pathogenesis of neuronal abnormalities and developing new therapeutic strategies for neurodevelopmental disorders.

I would like to dedicate this book to the families of individuals with intellectual disability. While they experience incredible pain and suffering, many do not wait for others to act. During my many years of work on neurodevelopmental disorders, I have come to the conclusion that without the contribution and active participation by parents, brothers and sisters, and other members of immediate family of individuals with intellectual disability, it would not have been possible to get where we are now. While developmental disabilities are not considered a top priority for funding and research by the governments in many regions of the world, the last decade, has witnessed multiple examples in which a very few hard-working and persistent family members of individuals with intellectual disability have made an enormous contribution in direct involvement and persuading others to get involved. This has indeed made a difference in fundamental research, clinical care, and developing treatments for many aspects of neurodevelopmental disorders. Dedicating this book to these family members is to thank them for believing that, once the scientists are equipped with knowledge on the molecular mechanisms of brain function and with the proper tools to modify the brain circuits involved in cognition, they can make a huge difference in all aspects of the lives of people with intellectual disability.

Dr. Ahmad Salehi
Clinical Associate Professor
Department of Psychiatry & Behavioral Sciences
Stanford Medical School
USA

Chapter 1

Cytogenomic Abnormalities and Dosage-Sensitive Mechanisms for Intellectual and Developmental Disabilities

Fang Xu and Peining Li

Additional information is available at the end of the chapter

http://dx.doi.org/10.5772/55689

1. Introduction

Within each human cell, double strand DNA molecules are packed into 22 pairs of autosomes numbered from 1 to 22 and one pair of sex chromosomes denoted as XX for female and XY for male. Every human chromosome has a centromere to guide its segregation through cell cycles and a telomere at each end to protect its integrity. Chromosomes play important roles in gene expression regulation, DNA replication and cell division. Abnormalities involving the number and the structure of each chromosome or a segment within a chromosome are known to introduce functional disturbance and cause genetic diseases.

Medical genetics has been driven by evolving technologic innovations for better genetic diagnosis and expanding clinical evidence for rational genetic counseling and disease treatment. Since 1970s, a series of technologies operating on differentiate staining of metaphase chromosomes or locus-specific hybridization of labeled DNA probes has been developed to study chromosomal and submicroscopic abnormalities. Karyotyping using Giemsa-stained banding pattern (G-band) on treated metaphase chromosomes and fluorescent in situ hybridization (FISH) mapping on metaphase chromosome or interphase chromatin are the standard procedures in clinical molecular cytogenetic laboratories. Molecular cytogenetic analysis of pediatric patients with developmental delay (DD), intellectual disability (ID), multiple congenital anomalies (MCA) and autistic spectrum disorders (ASD) has found many causative chromosomal abnormalities and some genomic disorders. In the past decade, genomic analysis using either oligonucleotide array comparative genomic hybridization (aCGH) or single nucleotide polymorphism (SNP) chip has been validated and recommended as the first-tier genetic testing for pediatric patients. This integrated cytogenomic approach further defines the genomic coordinates and gene content of chromosomal and cryptic genomic abnormalities and extends the spectrum of etiologic causes for ID/DD/MCA and ASD. The genomic information facilitates fine-mapping of disease-causing genes and dissecting underlying pathogenic mechanisms through *in silico* bioinformatic data mining, *in vitro* cellular phenotyping and *in vivo* animal modeling. Ultimately, this progress will lead to rational disease classification and therapeutic interventions for patients with ID/DD/ASD [1-3].

2. Cytogenetic and genomic methodologies

2.1. Molecular cytogenetic approach

Clinical cytogenetics is the study of human chromosomal abnormalities and their associated phenotypes. In 1956, Tjio and Levan [4] correctly described that a normal human metaphase contains 46 chromosomes. This fundamental cytologic observation was built upon the development of *in vitro* cell culture techniques along with the use of colchicine to arrest the cell cycle at metaphase and the modification of Hsu's hypotonic treatment prior to fixation to spread out the chromosomes [5]. The analysis of directly Giemsa-stained chromosomes led to the identification of numerical chromosomal abnormalities like trisomy 21 in Down syndrome [6], 45,X in Turner syndrome [7], 47,XXY in Klinefelter syndrome [8], trisomy 13 [9], and trisomy 18 [10]. The early discoveries of these syndromic numerical chromosomal abnormalities prompted efforts to differentiate all 23 pairs of chromosomes to detect structural abnormalities. In 1968, Caspersson et al. [11] reported differentiate Quinacrine staining of chromosomes and triggered the development of various chromosome banding techniques. Giemsa staining on trypsin-treated chromosome spreads forms unique Giemsa-positive and negative bands which looks like G-band 'barcodes' for each pair of chromosomes under the microscope. A normal human G-band ideogram was used as a standard for accurate grouping, numbering and pairing of human chromosomes based on their size, centromere position, defined regions and bands; this organized chromosomal profile of an individual is referred to as a karyotype [12]. Chromosome heteromorphisms mainly involving highly repetitive sequences in the pericentric and satellite regions have been recognized through studies on normal human populations and diagnostic practices [13]. General consensus on heteromorphic regions and their reporting practice was reported [14]. Despite an effective tool to detect numerical and structural chromosomal abnormalities, the banding method has two obvious technical limitations: the requirement of viable cells for setting up cell culture to capture metaphases for microscopic analysis and the low analytical resolution of chromosomal G-bands. The size of a human genome is 3000 Mb (megabases) and estimated total number of protein-coding genes is about 20,000. So the average size of a chromosome G-band in a medium 500-band level is about 6 Mb and contains 40 coding genes. Before the application of genomic technologies, the lack of genomic mapping for involved genes of many detected chromosomal abnormalities had been the major obstacles for accurate karyotype-phenotype correlation and candidate gene identification.

In 1982, FISH technology using labeled DNA probes hybridized onto metaphase chromosomes was developed to map genes onto specific chromosomal G-band regions [15]. This gene mapping tool was immediately recognized to have great diagnostic value. FISH on metaphase chromosomes, using labeled DNA probes in the size of 100-800 kilobase (Kb), has enhanced the analytical resolution and allowed accurate diagnosis of genomic disorders (also termed contiguous gene syndromes or microdeletion disorders), such as DiGeorge syndrome (OMIM#188400) by a deletion at 22q11.2, Prader-Willi syndrome (OMIM#176270) and Angelman syndrome (OMIM#105830) by a deletion at 15q11.2. FISH can also be performed directly on interphase nuclei, which overcame the limitation of cell culture and extended its diagnostic application toward rapid screening of chromosomal and genomic abnormalities. Multiplex FISH panels with differentially labeled probes have been developed for prenatal screening of common aneuploidies involving chromosomes X, Y, 13, 18 and 21 [16] and for postnatal detection of cryptic subtelomeric rearrangements [17].

The molecular cytogenetic approach combining G-banding and FISH technologies has been the standard for a primitive genome-wide view or a locus-specific view of numerical and structural chromosomal abnormalities. An international system for human cytogenetic nomenclature (ISCN) was first introduced in 1978 and has been continuously updated to the current 2013 version for a systematic documentation of chromosomal and genomic abnormalities [12]. Practice guidelines for cytogenetic evaluation of DD/ID/MCA have been established; the abnormality detection rate is 3.7% by conventional karyotyping for large numerical and structural chromosomal abnormalities and up to 6.8% when combined with FISH analysis for targeted genomic disorders and subtelomeric rearrangements [18].

2.2. Genomic analysis as first-tier genetic testing

In 1992, to overcome frequent cell culture failure and poor metaphase quality in karyotyping solid tumor samples, Kallioniemi et al. [19] developed comparative genomic hybridization (CGH) using differently labeled test and control DNAs co-hybridized onto normal metaphase chromosomes to measure copy number changes. In 1995, Schena et al. [20] developed a microarray-based technology to quantitatively monitor multiple gene expression. A hybrid of these CGH and microarray technologies formed the novel array CGH (aCGH) technology for a high resolution analysis of copy number changes through the genome. From 1998 to 2001, prototype CGH arrays with increased coverage from a single chromosome to the whole genome using spotted BAC or PAC clones were produced by academic laboratories [21, 22]. Five years later, high-throughput, high density oligonucleotide microarrays or single nucleotide polymorphism (SNP) chips following industrial standards along with user-friendly analytical software packages were developed by several companies. These novel genomic technologies quickly filled the gap between the megabase (Mb)-range chromosome G/R-bands and kilobase (Kb)-level gene structure and led to the discovery of polymorphic copy number variants (CNV) in the normal human genome [23, 24]. CNVs are defined as the gain or loss of genomic materials larger than 1 kb in size and they present in approximately 12% of the genome from normal human populations [24]. Meanwhile, these genomic technologies had also been applied to delineate chromosomal abnormalities and detect pathogenic genomic imbalances for patients with DD/ID/MCA in a research setting [25-28]. These technical and research progresses set a solid foundation for diagnostic application.

To ensure the safety and effectiveness of a new technology for genetic diagnosis, analytical and clinical validities followed by evidence-based practice guidelines have to be established. Genomic analysis involves multi-step bench procedures of DNA extraction, enzymatic labeling or extension and DNA hybridization, a large amount of data analysis and knowledge-based result interpretation. The integration of this DNA-based genomic analysis into a cell-based microscopic analysis could be a technical challenge for many clinical cytogenetic laboratories.

For genomic analysis, analytical validity refers to the probability that a test will be positive when particular copy number variants (deletion or duplication) are present (analytical sensitivity), the probability that the test will be negative when these variants are absent (analytical specificity), and the analytical resolution [29]. Most analytical validation studies compared the outcomes between the genomic analysis and the conventional cytogenetic method or among different platforms. Two earlier pilot studies compared array results with known chromosomal abnormalities from 25 cases to validate targeted BAC clone arrays, and the clone-by-clone sensitivity and specificity were estimated to be 96.7% and 99.1% respectively [30, 31]. Using a receiver operating characteristic (ROC) curve, the analytical validity of a genome-wide oligonucleotide aCGH (Agilent 44K) showed 99% sensitivity and 99% specificity when the analytical resolution was set at 300-500 Kb by five to seven contiguous oligonucleotides (about six times the average spatial resolution of 68 Kb, given by the coverage of a 3,000,000 Kb human genome with approximately 44,000 oligonucleotide probes). For the detection of mosaicism using the set resolution, aCGH can achieve 85% sensitivity and 95% specificity for a mosaic pattern at 50% of the cell population, but increased test-to-test variations and reduced sensitivity are expected as the mosaic percentage decreases [32]. Another validation study recommended similar analytical parameters by using a sliding window of four to five oligonucleotide probes [33]. Additionally, the comparison of the area under the ROC curve clearly demonstrated that the analytical validity of oligonucleotide aCGH outperformed BAC clone aCGH [32]. The superior performance of oligonucleotide aCGH over BAC clone aCGH was later confirmed in a comparative analysis [34]. ROC analysis is effective in evaluating and comparing analytical validity among different technical platforms for a rational decision in selecting a novel technology for diagnostic application.

Cross platform comparison on a 33K tiling path BAC array, 500K affymetrix SNP chip, 385K Nimblegen oligonucleotide array and 244K Agilent oligonucleotide array was performed using ten cases with known genomic imbalances ranging from 100 Kb to 3 Mb. Sensitive performances were noted in all platforms, but accurate and user-friendly computer programs are of crucial importance for reliable copy number detection [35]. Technically, one obvious advantage of the SNP chips over aCGH is the ability to detect uniparental disomy (UPD), copy neutral loss of heterozygosity (CN-LOH) or absence of heterozygosity (AOH), and level of consanguinity and incest [36]. However, the introduction of SNP probes into CGH array by Agilent Technologies has resolved the technical differences to a certain extent. Validation studies of UPD detection using the Affymetrix SNP genechips detected isodisomic UPD and segmental AOH of a defined size but missed heterodisomic UPD [37, 38]. The high density aCGH and SNP chips have achieved exon-by-exon coverage to detect intragenic and exonic copy number changes, which could allow direct evaluation of genotype-phenotype concordance [39]. A recent study using exon-level high density aCGH on a targeted list of genes showed the detection of mostly exonic deletions in 2.9% cases with autosomal dominant disorders, intragenic deletions in 10.1% of cases of autosomal recessive disorders tested with one known mutation, and a deletion and duplication in 3.5% of X-linked disorders [40]. Laboratories pursuing this high resolution genomic analysis will require additional validation studies, in-depth result interpretation into Mendelian disorders, more familial follow up studies for incidental or secondary findings, and eventually further functional studies.

Clinical validation refers to the probability that a test will be positive in people with the disease (clinical sensitivity) or negative in people without the disease (clinical specificity) [29]. There were concerns about false negative results, procedure variability and interpretation criteria for the clinical application of early versions of targeted BAC clone array [41]. Due to the high cost of aCGH and SNP chip analysis, a validation study on a large number of patient and control samples in every clinical laboratory is not practical. The ACMG guidelines recommend a validation procedure of analyzing a minimum of 30 specimens with different known chromosomal abnormalities [42]. Most cytogenetic laboratories performed a parallel comparison between aCGH or SNP chip analysis and karyotyping on a small case series. All these studies demonstrated consistently that the aCGH or SNP chip can define the genomic coordinates and gene content of chromosomally observed imbalances and also detect cryptic microdeletions, microduplications and subtelomeric rearrangements. For example, a focus oligonucleotide aCGH was validated with 100% concordance toward known chromosomal imbalances and yielded an 11.9% abnormal detection rate of 211 clinical samples [43]. Oligonucleotide aCGH using high density Agilent 244K was validated in 45 cases with known chromosomal abnormalities and microdeletions [44]. A multi-center comparison of 1,499 patients using the same oligonucleotide platform (Agilent 44K) showed a 12% abnormality detection rate, and about 53% of the abnormal findings are less than 5 Mb and thus beyond the analytical resolution of routine karyotyping [45]. The clinical validity of genomic analysis could be estimated from data published by the International Standards for Cytogenomic Arrays consortium (ISCA). Using 14 well known recurrent microdeletions and microduplications ranging from 1.5-3 Mb as a reference set of genomic disorders, 458 microdeletions and 270 microduplications were detected from 15,749 pediatric patients and 12 microdeletions and 53 microduplications were found from 10,118 published controls [46]. Given the analytical sensitivity of 99% at a resolution of 300-500 Kb for a routine aCGH (Agilent 44K) [32], the clinical sensitivity is close to 100% for this reference set of genomic disorders; given the 65 false negative results from 10,118 controls, the clinical specificity is estimated to be 99.4%. The near-perfect validities and significantly improved resolution of genomic technologies made them ideal diagnostic tools for delineating chromosomal imbalances and detecting CNVs.

2.3. Integrated cytogenomic workflow and practice guidelines

The current practice guidelines from the American College of Medical Genetics (ACMG) and the peer consensus recommend that genomic analysis be the first-tier genetic testing for pediatric patients with DD/ID/MCA/ASD [47-50]. A CNV detected from a normal individual is likely a polymorphic variant without clinical significance and usually termed as a benign CNV (bCNV). A CNV with known disease association is referred as pathogenic CNV (pCNV), and a rare or private CNV with uncertain clinical relevance is named a variant of unknown significance (VUS) [51]. For the past five years, clinical genomic analysis has progressed rapidly from 'targeted' or 'focused' aCGH to genome-wide high resolution aCGH or SNP chips. A systematic review of 21,698 pediatric patients analyzed by different genomic platforms demonstrated a diagnostic yield of 15-20% [47], which is two to three times higher than the 3.7-6.8% yield by molecular cytogenetic analysis [18].

In diagnostic practice, there are serious concerns regarding the complete replacement of molecular cytogenetic approach. The analytical validities and technical capacity of chromosome, FISH and oligonucleotide aCGH are summarized in Table 1. Although cytogenetic testing has gradually become a supplemental or confirmatory procedure, karyotyping is still the gold standard to detect numerical chromosomal abnormalities (e.g., trisomy 21 for Down syndrome and monosomy X for Turner syndrome) and structural rearrangements (e.g. Robertsonian translocation) and FISH is also the 'cell-based' method of choice to determine mosaic patterns. Approximately 45% of genomic imbalances are larger than 3-5 Mb and could be confirmed by high resolution G-banding; most recurrent genomic disorders, subtelomeric rearrangements and mosaic patterns can be readily confirmed by clinically-validated commercial FISH probes [45].

In a recent study on the interpretation and reporting of CNVs without known associated abnormal phenotype from different laboratories, it was found that none of the thirteen CNVs was in complete agreement and the interpretations ranged from normal to abnormal in some cases [52]. In fact, some genomic findings will always be difficult to interpret because of their variable expressivity and incomplete penetrance. The collection of bCNVs, pCNVs and VUSs into web-delivered databases has provided an essential tool in interpreting results for diagnostic laboratories and also in educating clinical geneticists and genetic counselors [53]. The websites of Database of Genomic Variants (DGV), International Standards for Cytogenomic Arrays (ISCA), DatabasE of Chromosomal Imbalance and Phenotype in Humans using Ensembl Resources (DECIPHER) and other related web resources are listed at the end of this report. CNVs in the DGV, DECIPHER and ISCA have been loaded onto the Human Genome Browser as searchable tracks. The brief clinical description from DECIPHER [54] and the evidence-based rating of CNV into pathogenic, likely pathogenic, uncertain significance, likely benign and benign from ISCA [55] are all helpful in reporting genomic findings. Recognized cytogenomic syndromes usually have entries in the Online Mendelian Inheritance in Man (OMIM). Novel cytogenomic abnormalities are usually presented as case reports and can be search from PubMed. Reports of a series of similar findings and in-depth reviews will provide more evidence. Figure 1 shows an aCGH detected 16p13.11 deletion in a patient with ASD in comparison with pathogenic deletions and duplications documented in the DECIPHER and ISCA databases. Searchable clinical information from these databases could be used to assure genotype-phenotype correlations. According to the genotype-phenotype correlation, a four-level evidence-based result interpretation scheme has been proposed [56]:

Level I: Tight genotype-phenotype correlation with well-defined association of the described syndrome and the pCNV. There may be variability in the phenotype, but the spectrum of variability is well described (e.g. **Williams syndrome** and **DiGeorge syndrome**). Many known syndromes have an assigned OMIM number which could be used directly as a reference in the report.

	Analytical Validity*				Types of Abnormalities Detected**						
	Spatial	Analytical			Chr	Bal	Unbal		UPD		
	Resolution	Resolution	Sensitivity	Specificity	Num Abn	Struc Abn	Struc Abn	CNV	AOH	Exonic	Mosaic
Cell-based											
G-banding											
Routine (400-550 bands)		5 ~ 7 Mb			+	+	+	-	-	-	>6%
High Resolution (550-850 bands)		3 ~ 5 Mb			+	+	+	-	-	-	>6%
FISH											
Gene/locus-specific		100-800 Kb	~98%	~98%	+	+	+	+	-	-	>3~5%
Regional specific (cen/subtel)		>100 Kb	~98%	~98%	+	+	+	+	-	-	>3~5%
DNA-Based											
Oligonucleotide aCGH (Agilent)											
Human CGH 44K	68 Kb	400~500 Kb	>99%	>99%	+	-	+	+	-	-	>20%
Human CGH 180K	17 Kb	100~120 Kb	>99%	>99%	+	-	+	+	-	-	>20%
Human CGH+SNP 180K	17 Kb	100~120 Kb	>99%	>99%	+	-	+	+	+	-	>20%
Human CGH+SNP 400K	7.5 Kb	40~50 Kb			+	-	+	+	+	+	>20%

* Sensitivity and specificity of FISH based on laboratory validation and of aCGH based on ref. #32

** Chr = chromosome, num = numerical, bal = balanced, unbal = unbalanced, Abn = abnormalities; CNV = copy number variant, UPD = uniparent disomy, AOH = Absence of heterozygosity, + detectable, - undetectable; percentage of mosaic detection based on chromosome analysis of 50 metaphases, FISH assay of 200 cells and ref#32 for aCGH.

Table 1. Analytical Validities and Diagnostic Capacity of Cytogenomic Analyses

Level II: Evolving genotype-phenotype correlation. The described syndrome is represented by a case report in the literature, or is associated with more than one distinct phenotype and may be influenced by penetrance or modifying genes, such as **1q21.1 duplication or deletion syndrome**.

Level III: Possible new genotype-phenotype correlation. The CNV has not been described in the literature before, and no published phenotypic data are available.

Level IV: Uninterpretable multiple CNVs and VUSs. None of which are reported as normal variants or associated with a disease phenotype, or one or more of the CNVs/VUSs are also found in a phenotypically discordant parent.

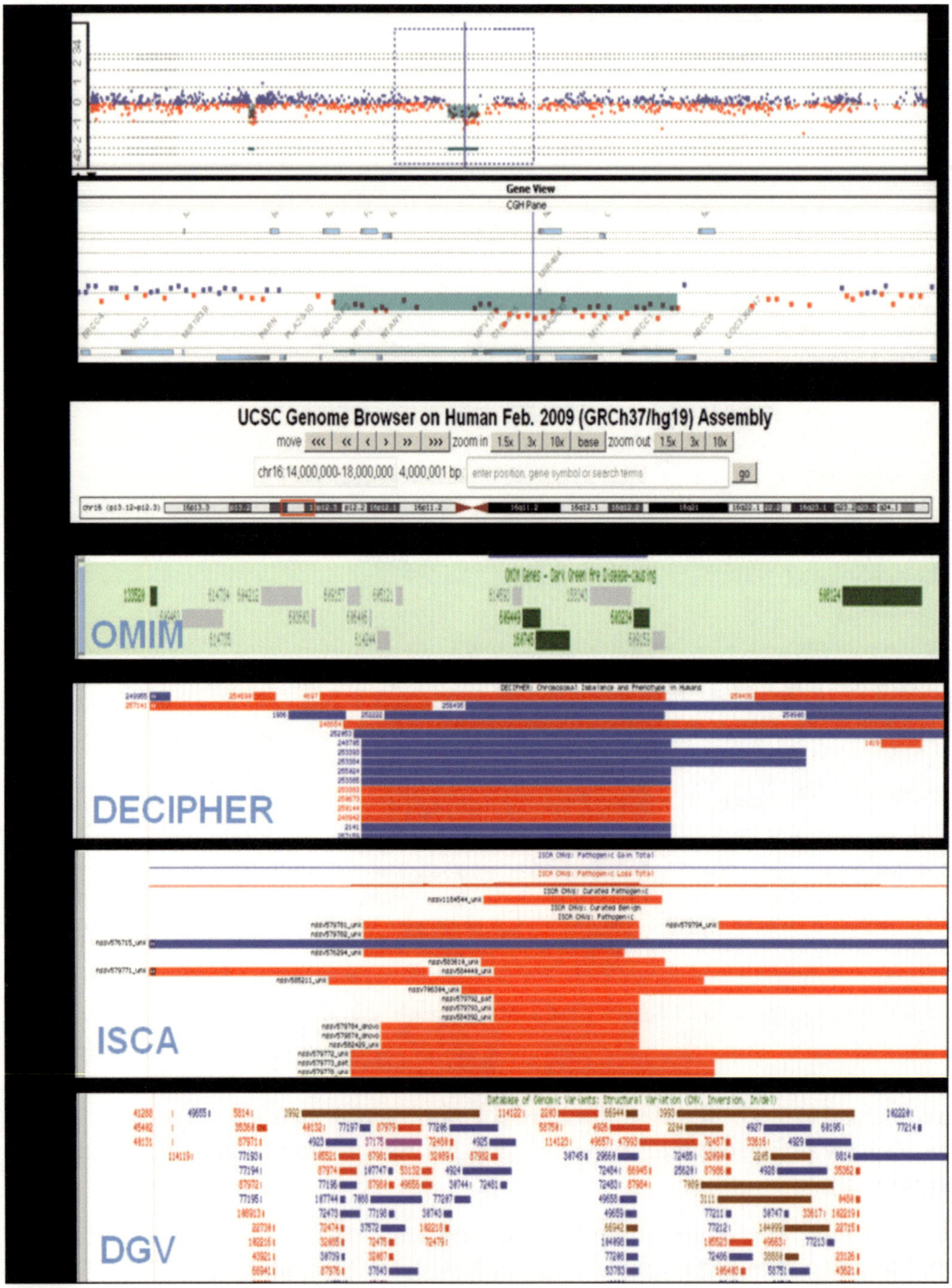

Figure 1. Diagnostic interpretation using web-delivered databases. A. A 16p13.11 deletion detected by Agilent's 180K CGH+SNP array. B. The deleted region is shown in the UCSC genome browser with searchable tracks of OMIM genes, DECIPHER, ISCA and DGV.

Each laboratory that performs genomic analysis should develop its quality control and quality assurance procedures. Proficiency testing for genomic analysis has been implemented by the College of American Pathologists (CAP). Since 2008, two pilot and ten survey challenges of twelve DNA specimens distributed to as many as 74 different laboratories yielded 493 individual responses with a 95.7% mean consensus for matching result interpretations. Responses to supplemental questions indicate that 72% of laboratories use oligonucleotide aCGH and 23% use SNP chips, and array platforms used are increasing in probe density [57].

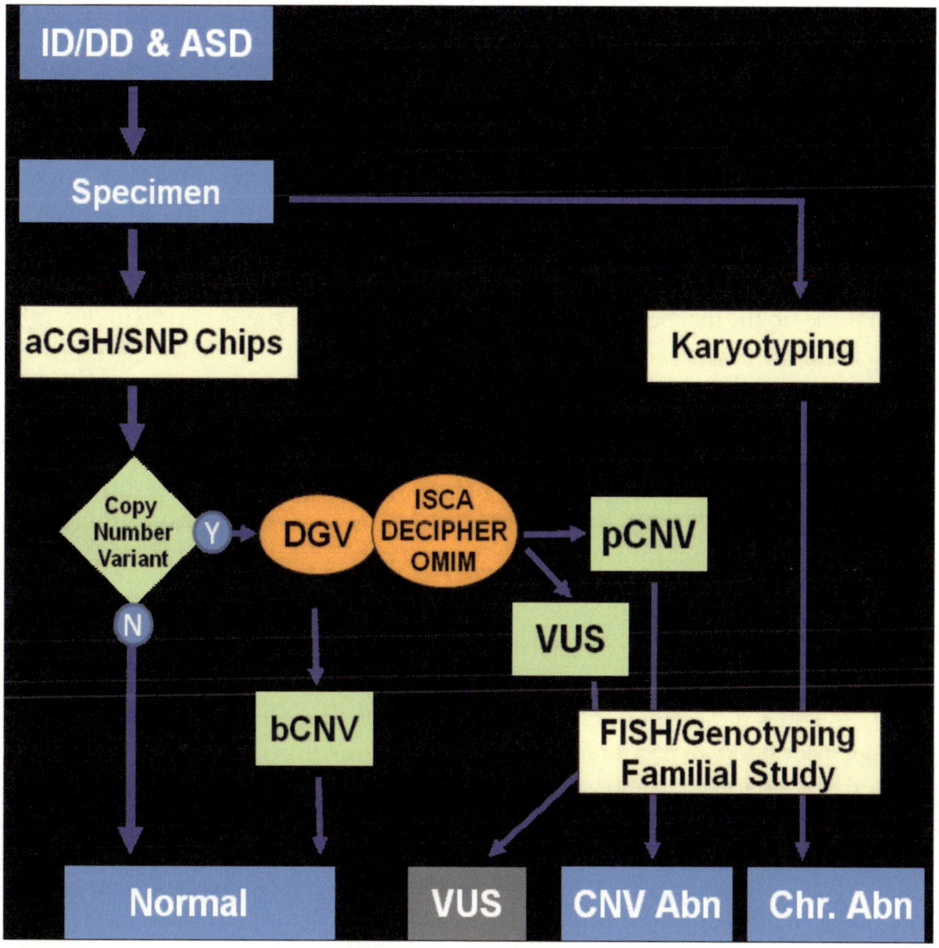

Figure 2. A workflow illustrates the integrated cytogenomic approach for pediatric genetic evaluation. Patients with ID/DD/ASD are tested by aCGH first and limited karyotyping. Detected copy number variants are assessed using web-delivered databases DGV, ISCA, DECIPHER and OMIM to define pCNV, bCNV and VUS. Follow-up confirmatory FISH and genotyping and familial study to determine parental origin of VUS and pCNV are considered.

Taken together, the development of practice guidelines and proficiency testing indicated that the aCGH and SNP chip analyses are becoming the 'gold standard' in clinical diagnosis of chromosomal and genomic abnormalities. As shown in Figure 2, a workflow integrating the cell-based molecular cytogenetic methods (G-band and FISH), the DNA-based genomic copy number detection (aCGH or SNP chip) and evidence-based result interpretation has been the most efficient and cost-effective diagnostic cytogenomic setting. On the clinical front, pediatric genetic evaluation should be arranged with pre- and post-test consultations by a medical geneticist so that the benefits, limitations, and possible outcomes, as well as the difficulties of interpreting some copy number variants can be discussed in details.

3. Spectrum of cytogenomic abnormalities in ID/DD and ASD

The prevalence of ID/DD and ASD are reported to be 1~3% and 0.67%, respectively [58]. Other common neurodevelopment disorders including speech and language delay, schizophrenia and epilepsy are also subjected for cytogenomic testing. The integrated cytogenomic analysis has significantly improved the diagnostic yield from 3.7-6.8% by molecular cytogenetic analysis [18] to 15-20% by oligonucleotide aCGH or SNP gene chip [47]. The spectrum of cytogenomic abnormalities ranges from large interstitial and subtelomeric imbalances, submicroscopic recurrent genomic disorders, to cryptic oligo-genic to intragenic copy number changes. However, the diagnostic yield could be varied by the criteria of patient referrals and the resolution of genomic analysis. For example, from 2006-2011, there were 1,354 consecutive pediatric patients analyzed by 44K and 180K oligonucleotide aCGH (Agilent) in Yale Cytogenetic Laboratory and pathogenic abnormalities were detected in 176 patients (a 13% diagnostic yield). These abnormalities were classified into chromosomal and cryptic structural abnormalities 95 patients (54%, 95/176), recurrent genomic disorders in 66 patients (37.5%, 66/176), and common aneuploidies in 15 patients (8.5%).

3.1. Chromosomal and cryptic structure abnormalities

With its much higher analytical resolution than chromosome G-banding, genomic analysis can delineate the genomic coordinates and gene contents for almost all chromosomally visible numerical and structural imbalances. This genomic information facilitates fine mapping of critical regions or intervals containing candidate dosage-sensitive genes through subtractive comparison of overlapped deletions and duplications [59-61]. In most recent case reports, the critical regions defined by aCGH or SNP chip have clearly mapped genomic coordinates and accurately defined gene content. As more cases involving similar genomic locus accumulated, the mapping of a critical region can be narrowed down to a few candidate genes or even a single gene. A typical example is the mapping of dosage-sensitive genes associated with microcephaly, corpus callosum agenesis and seizure from chromosome 1q43-q44 deletion syndrome (OMIM#612337). This syndrome is caused by heterogeneous subtelomeric deletions of 1q43-q44. Based on 10 cases of unrelated patients with 1q43-q44 subtelomeric deletions, Van Bon et al. [62] defined a 360 Kb critical region of 1q44 with four candidate genes C1orf100, ADSS, C10rf101 and C1orf121 for corpus callosum abnormality. A series of studies including cases with small interstitial deletions suggested that the nearby genes AKT3, ZNF238 and HNRNPU are more likely the candidate genes [63-67]. A de novo 163 Kb interstitial microdeletion at 1q44 involving only the HNRNPU and FAM36B genes was reported in a boy with thin corpus callosum, psychomotor delay and seizure [68]. Combined data from these studies supported three critical regions containing AKT3, ZNF238 and HNRNPU genes for microcephaly, corpus callosum abnormalities and seizure, respectively. However, the term "corpus callosum abnormalities" associated with 1q44 deletions include a spectrum of developmental aberrations from complete agenesis, partial agenesis, dysgenesis, hypoplasia and thin corpus callosum. The modifying effect from gene interaction or genetic background could also contribute to the phenotype. Further experimental study on gene function and interaction is needed to fully understand the genotype-phenotype correlation.

Genomic analysis can also resolve the genomic structures, mutagenesis mechanisms and mitotic or meiotic behaviors from puzzling chromosomal structural abnormalities like ring chromosomes or supernumerary marker chromosomes [69-71]. For example, ring chromosome 20 syndrome is a rare chromosomal disorder characterized by refractory epilepsy with seizures in wakefulness and sleep, behavior problems and mild to severe cognitive impairment. The aCGH analysis revealed two distinct groups of patients: 75% were mosaic for the r[20] and a normal cell line with no detectable deletions or duplications of chromosome 20 in either cell line and 25% had non-mosaic ring chromosomes with a deletion at one or both ends of the chromosome. The age of onset of seizures inversely correlated with the percentage of cells containing the ring chromosome [72]. Another interesting observation from aCGH applications on two large series is the detection of low-level mosaicism of numerical and structural abnormalities in approximately 0.5% of patients referred for DD/ID/MCA [73, 74]. The authors suggested that the DNA extracted from the white blood cells can reflect mosaic pattern more accurately than culture stimulated lymphocytes. A cytogenomic approach combining cell-based methods of FISH on directly prepared interphase cells and extensive karyotyping on metaphase cells with DNA-based estimation from aCGH log2 ratio or SNP pattern was proposed for dissecting mosaic patterns [70].

Hidden genomic aberrations in complex chromosomal rearrangements or apparently balanced translocations were also detected by aCGH [75, 76]. Of patients presenting abnormal phenotypes and an apparently balanced translocation, approximately 29-40% has cryptic breakpoint-associated or unrelated imbalances of paternal origin [77, 78]. Several disease-causing mechanisms induced by a balanced translocation including loss of function by gene disruption, gain of function by gene fusion and aberrant expression by positional effect have been demonstrated. For example, Cacciagli et al. [79] detected a de novo balanced translocation t(10;13)(p12;q12) in a patient with severe speech delay and major hypotonia. This translocation disrupted the ATP8A2 gene. This gene is highly expressed in the brain, suggesting the patient's mental disability is likely due to the halpoinsufficiency of the ATP8A2 gene. Brownstein et al. [80] reported a case with over-expression of the α-Klotho gene induced by a balanced translocation t(9;13)(q21.13;q13.1) and established the association α-Klotho over-expression with hypophosphatemic rickets and hyperparathyroidism. Application of paired-end genomic sequencing or breakpoint-targeted capture sequencing on five ASD/DD patients carrying a balanced rearrangement revealed unexpected sequence complexity as an underlying feature of karyotyping balanced alterations [81]. Cost-effective diagnostic sequencing analysis for balanced rearrangements detected in patients with ID/DD/ASD should be implemented in the near future.

3.2. Recurrent genomic disorders

Genomic disorders refer to microdeletions and microduplications mediated by non-allelic homologous recombination (NAHR) within regional low copy repeats (LCRs). A dozen of recurrent genomic disorders such as DiGeorge syndrome caused by a deletion at 22q11.2, Williams-Beuren syndrome (OMIM#194050) by a deletion as 7q11.23, Prader-Willi syndrome and Angelman syndrome by a deletion at 15q11.2 have been recognized clinically and routinely diagnosed by FISH testing. The application of genomic analysis enables not only more accurate diagnosis of these previously recognized genomic disorders but also the detection of many novel recurrent genomic disorders. In 2006, the first genomic disorder identified by aCGH was a 500 Kb microdeletion at 17q21.31 containing the MAPT gene (microtubular associated protein tau) from patients with a clearly recognizable ID, hypotonia and a characteristic face [82, 83]. This later termed Koolen syndrome (OMIM#610443) is caused either by heterozygous mutation in the KANSL1 gene or a 17q21.31 deletion. The KANSL1 gene encodes a nuclear protein that plays a role in chromatin modification. It is a member of histone acetyltransferase (HAT) complex. The reciprocal 17q21.31 microduplication syndrome (OMIM#613533) manifestsing some degree of psychomotor retardation and poor social interaction and communication difficulties reminiscent of ASD was reported [84]. Since then, many genomic disorders have been reported, and the diagnosis of these genomic disorders has become an integral part of pediatric genetic evaluation. The aCGH analysis on 15,767 pediatric patients with ID/DD estimated that ~14.2% of them are caused by pCNVs over 400Kb, and approximately 60% of these pCNVs are within 45 known genomic disorder regions [85]. An evidence-based approach was used to establish the functional and clinical significance of the most commonly seen 14 genomic disorders [46]. Table 2 lists the recognized dosage-sensitive genes and estimated penetrance, frequency and prevalence of these 14 recurrent genomic disorders. These most frequently seen genomic disorders represent a 4.5% (1/22) diagnostic yield in pediatric patients and an estimated 0.18% (1/550) prevalence in a general population. A study of human populations for the polymorphic inversions at 17q21.31 observed that the H2 haplotype occurred at the highest frequencies in South Asian and Southern Europe; this H2 haplotype is susceptible to de novo deletions that lead to developmental delay and learning difficulties [86]. Population genetic studies for genomic disorders of other loci could define predisposing genomic structures and recurrence risk for different ethnic groups at different geographic regions.

The microdeletion and microduplication of the same genomic locus offer an opportunity to study dosage-sensitive genes, especially for the opposite phenotypes of haploinsufficient and triple-sensitive genes. Although clinical evaluation has been complicated by overlapped phenotypes, variable expressivity and reduce penetrance for many newly-defined genomic disorders, opposite phenotypes have been seen in a few genomic disorders. Comparison of clinical features of 7q11.23 microdeletion for Williams syndrome with reciprocal microduplication syndrome (OMIM#609757) noted different neurologic and behavior problems. The

Chromosome G-band	Genomic locus (Mb in hg18)	Dosage-sensitive genes	Abn.*	Clinical features (OMIM#)*	Penetrance**	Ref #46 (15,749 cases)	Ref #85 (15,767 cases)	Frequency (31,516)†	Est. prevalence‡
22q11.2	17.4-18.7	*TBX1*	del	DiGeorge syndorme (188400)	1	93	96	1/167	1 in 4,000
			dup	22q11.2 duplication syndrome (608363)	0.91	32	50	1/384	1 in 9,200
16p11.2	29.5-30.1	*TBX6, KCTD13*	del	Autism, Obsity (611913)	0.96	67	64	1/241	1 in 5,800
			dup	Autism, underweight	0.93	39	28	1/470	1 in 11,300
1q21.1	145.0-146.4	*HYDIN2*	del	1q21.1 deletion syndrome (612474)	0.96	55	47	1/309	1 in 7,400
			dup	1q21.1 duplication syndrome (612475)	0.96	28	26	1/584	1 in 14,000
15q13.2-q13.3	28.7-30.3	*CHRNA7*	del	15q13.3 microdeletion syndrome (612001)	1	46	42	1/358	1 in 8,600
			dup	Psychiatric disease	0.87	14	20	1/927	1 in 22,200
7q11.23	72.2-77.5	*ELN, GTF21*	del	Williams syndrome (194050)	1	34	42	1/414	1 in 10,000
		FZD9, LIMK1	dup	7q11.23 duplication syndrome (609757)	1	16	16	1/985	1 in 23,600
15q11.2-q13	22.3-26.1	*GABRA5*	del	Prader-Willi (176270)/Angelman syndromes (105830)	1	41	16	1/552	1 in 13,300
			dup	15q11-q13 duplication syndrome (608636)	1	35	27	1/508	1 in 12,200
17q21.31	41.0-41.7	*MAPT, KANSL1*	del	17q21.31 deletion syndrome (610443)	1	22	23	1/700	1 in 16,800
			dup	17q21.31 duplication syndrome (613533)	0.43	21	2	1/1,370	1 in 32,900
16p13.11	15.4-16.2	*MYH11*	del	Autism, ID, and schizophrenia	0.86	22	18	1/788	1 in 18,900
			dup	Variable phenotype	0.71	45	24	1/457	1 in 11,000
17p11.2	16.6-20.4	*RAI1*	del	Smith-Magenis syndrome (182290)	1	16	16	1/985	1 in 23,600
			dup	Potocki-Lupski syndrome (610883)	1	15	9	1/1,313	1 in 31,500
17q12	31.9-33.3	*TCF2*	del	Renal cysts and diabetes (137920)	0.88	18	14	1/985	1 in 23,600
			dup	Epilepsy	0.86	21	18	1/808	1 in 19,400
1q21	144.1-144.5	*HFE2*	del	Thrombocytopenia-absent radius syndrome (274000)	0.87	17	13	1/1,050	1 in 25,200
			dup	1q21.2 duplication	0.81	9	25	1/927	1 in 22,200
8p23.1	8.1-11.8	*SOX7, CLDN23*	del	8p23.1 deletion syndrome	1	10	7	1/1,853	1 in 44,500
			dup	Variable phenotype	1	6	7	1/2,424	1 in 58,200
5q35	175.6-176.9	*NSD1*	del	Sotos syndrome (117550)	1	8	8	1/1,969	1 in 47,300
			dup	Short stature, microcephaly, speech delay	n/a	2	0	1/15,758	1 in 378,200
3q29	197.2-198.8	*DLG1*	del	3q29 microdeletion syndrome (609425)	1	9	6	1/2,101	1 in 50,400
			dup	3q29 duplication syndrome (611936)	1	8	4	1/2,626	1 in 63,000
Total								1 in 22	1 in 550

* Abn., abnormality; del, deletion; dup, duplication, OMIM#, Online Mendelian Inheritance in Man

** Penetrance of syndromic phenotypes was from Cooper et al. 2011

†Frequency of each genomic disorder was calculated by case numbers from kaminsky et al (ref.#46) and Cooper et al (Ref #85); the frequency of 14 genomic disorders is 1/22 of pediatric ID/DD/MCA/ASD patients.

‡ Prevalence was estimated by multiply frequency by 0.04 (using 4% for DD/ID/MCA/Autism in a general population); the prevalence of 14 genomic disorders is 1/550 in a general population

Table 2. Prevalence and penetrance of common genomic disorders from two large patient-control series

7q11.23 microdeletion shows relative strength in expressive language and excessive sociability. To the contrary, the 7q11.23 microduplication has speech and language delay, deficit of social interaction and aggressive behavior. The FZD9, LIMK1, CLIP2 and GTF21RD1 genes have been suggested to be the candidate genes for neurologic and behavior phenotypes [87]. Microdeletion syndrome at 16p11.2 (OMIM#611913) and microduplication syndrome at 16p11.2 (OMIM#614671) were initially associated with ASD [88] but a subsequent study revealed mirror body mass index phenotypes [89]. Microdeletion at 16p11.2 is often associated with obesity, macrocephaly and ASD, while reciprocal microduplication is associated with underweight, microcephaly and schizophrenia. Chromosome 1q21.1 microdeletion syndrome (OMIM#612474) and the reciprocal 1q21.1 microduplication syndrome (IMIM#612475) associated with microcephaly or macrocephaly and developmental and behavioral abnormalities, respectively [90-92]. Brunetti-Pierri et al. [91] suggested that the HYDIN paralog located in the 1q21 interval is a dosage-sensitive gene exclusively expressed in brain; HYDIN haploinsufficience is responsible for the microcephaly and HYDIN triple-sensitive effect is for the macrocephaly.

Reciprocal microdeletion and duplication could present distinct, overlapping or similar phenotype likely caused by the same dosage-sensitive gene. Hereditary neuropathy with liability to pressure palsies (HNPP) (OMIM#162500) is caused by a microdeletion of the PMP22 gene at 17p11.2 and Charcot-Marie-Tooth Neuropathy type 1A (OMIM#118220) is caused by a reciprocal microduplication. Both present demyelinating neuropathies but the phenotypes are distinct. Smith-Magensis syndrome is caused by a microdeletion at 17p11.2 involving the RAI1 gene and Potocki-Lupski syndrome is caused by the reciprocal microduplication. In human, both syndromes show variable mental retardation, motor and speech delay, sleep disturbance, behavior problems and autistic features. In mouse models, mice with RAI1 deletion show hypoactive, decreased anxiety and decreased dominance like behavior. To the contrary, mice with RAI1 duplication show hyperactive, increased anxiety and increased dominance behavior [93]. All these findings indicate the presence of different type of dosage-sensitive genes and the importance of careful and detailed clinical observations and functional understanding of multiple and complex dosage-sensitive mechanisms.

3.3. UPD (uniparental disomy), AOH (absence of heterozygosity) and VUS (variants of unknown significance)

UPD is defined as the inheritance of both homologs of a chromosome pair from a single parent. When both homologs are from that parent, it is denoted as heterodisomy or heteroUPD. If both copies are from one parental homolog, it is termed as isodisomy or isoUPD. UPD of chromosomes 6, 7, 11, 14 and 15 have been known to cause diseases. Paternal UPD of chromosome 15, patUPD15, causes Prader-Willi syndrome while maternal UPD, matUPD15 causes Angelman syndrome. Segmental duplication of maternal 11p15 or paternal deletion of 11p15 causes decreased expression of IGF2, manifesting with impaired growth for Silver–Russell syndrome. Segmental duplication of paternal 11p15, paternal UPD, or maternal imprinting mutations of 11p15 lead to increased expression of IGF2, manifesting with overgrowth for Beckwith–Wiedemann syndrome [94]. Current validated CGH-SNP aCGH and SNP chip can detect chromosomal and segmental isoUPD but the detection of heteroUPD requires concurrent parental study [36-38]. The clinical significance of AOH segments is not clear. One possible disease-causing mechanism could be the presence of autosomal recessive phenotype by the doubling of a single mutation within the AOH segment. Other findings such as VUSs detected in approximately 9.3% of pediatric cases will require follow up parental study to determine the parental origin of VUS and even further functional analysis to understand their clinical significance [46].

4. Mapping candidate genes and understanding dosage-sensitive mechanisms

The characterization of genomic coordinates of pathogenic chromosomal and genomic abnormalities in patients with ID/DD/MCA and ASD provides the opportunities to map dosage-sensitive genes. Genomic imprinting is a known dosage-sensitive mechanism regulating gene expression from only the paternal or maternal genes. By stringent definition, a dosage-sensitive gene will cause opposite phenotypes by haploinsufficiency in a deletion and triple-overdose in a duplication. However, ascertainment bias could be introduced due to assumed more severe phenotype in a deletion than a duplication. An accurate genotype-phenotype correlation will require systematic and replicate studies on large case-control series. For all newly-defined genomic disorders, more family-based and case-control pathophysiologic and disease course studies as well as functional studies in *in vitro* cellular models or *in vivo* animal models are needed. Functional characterization of dosage-sensitive mechanisms will provide better understanding of human development and rational intervention for ID/DD/MCA and ASD.

4.1. *In Silico* mining revealing interacting brain expressed gene from cytogenomic abnormalities

The development of high-throughput genomic techniques has prompted the introduction of numerous bioinformatic data mining tools to study gene function and interaction. It has been hypothesized that brain expressed genes and their interaction could play important roles in human mental development. *In Silico* bioinformatic analyses have been used to identify candidate genes and functional interactions from a small case series of pediatric patients and a large case-control study of ASD patients [95, 96]. It has been hypothesized that ID phenotype from different pCNVs and segmental imbalances may be caused by the functional disturbance of genes in interacting pathways or networks. To reveal candidate brain-expressed genes (BEGs) and their interacting networks from detected cytogenomic abnormalities, a discovery-driven *in silico* analysis using bioinformatic tools was performed [97]. Of the 1,354 patients analyzed by oligonucleotide array comparative genomic hybridization (aCGH) in a five-year interval in Yale cytogenetics lab, pathogenic abnormalities were detected in 176 patients, including recurrent genomic disorders in 66 patients, subtelomeric rearrangements in 45 patients, interstitial imbalances in 33 patients, chromosomal structural rearrangements in 17 patients and aneuploidies in 15 patients. Subtractive mapping of bCNV and extractive constructing of smallest overlapped regions defined 82 disjointed critical regions from the detected abnormalities. All genes from these critical regions were sorted by functional annotation using Database for Annotation, Visualization, and Integrated Discovery (DAVID) and by tissue expression pattern from Uniprot. A list of 461 BEGs generated from 73 disjointed critical regions was denoted. Enrichment of central nervous system specific genes in these regions was noted, and the number of BEGs increased with the size of the regions. Further gene prioritization using Gene Relationships Across Implicated Loci (GRAIL) identified candidate BEGs with significant cross region interrelation from data sources of PubMed abstract, gene ontology terms and expression pattern. Figure 3 shows the cross-loci interactions of gene ontology from detected deletions and duplications. This result implied shared cellular component and biological process from the defined candidate BEGs. Pathway analysis using Ingenuity Pathway Analysis (IPA) denoted five significant gene networks involving cell cycle, cell-to-cell signaling, cellular assembly, cell morphology, and gene expression regulations. Previous studies and our preliminary data support a model of polygenic interactions and multiple functional networks for human mental development. Further experimental study on the cellular function of these candidate genes and their interactions will lead to a better understanding of dosage-sensitive mechanisms.

4.2. *In vitro* cellular phenotyping and *in vivo* animal modeling

Little is known about the cellular and developmental functions of many newly identified dosage-sensitive genes. The understanding of dosage-sensitive mechanism for specific gene or pathway could lead to targeted treatment using protein inhibitor or small RNA interference. The limited availability and accessibility of live brain and neuron tissues is the major obstacle in the study of disease-causing mechanisms in human mental development. Recent progress in stem cell technologies has made possible the modeling of disease using patient derived stem cells. In 2010, Marchetto et al. [98] developed a culture system using induced pluripotent stem cells (iPSCs) from Rett syndrome patients' fibroblasts. These Rett syndrome iPSCs were able to undergo X-inactivation and generate functional neurons. Neurons derived from these iPSCs had fewer synapases, reduced spine density, smaller soma size, altered calcium signaling and electrophysiological defects when compared to controls. This cellular model provided critical evidence of an unexplored developmental window before disease onset and enable direct testing of drug effect in rescuing synaptic defects. Similarly, *Bona fide* iPCSs were generated from fibroblasts of a patient with Prader-Willi syndrome bearing 4p/15q translocation [99]. These iPSCs retained the DNA methylation in the imprinting center of maternal allele and reduced the expression of the disease-associated SNRPN gene, therefore, could be differentiated into neuron tissue to modeling cellular phenotype. Using skin fibroblasts from affected patients or amniocytes from prenatal diagnosis tests, iPCSs for monosomy X (Turner syndrome), trisomy 8 (Warkany syndrome 2), trisomy 13 (Patau syndrome) and partial trisomy 11;22 (Emanuel syndrome) were generated for further studies of global gene expression and tissue-specific differentiation [100]. All these reports indicated that stem cell technology offers reproducible *in vitro* cellular phenotypes for better understanding of neurodevelopment and also a testable system for the development of therapeutical approach [101].

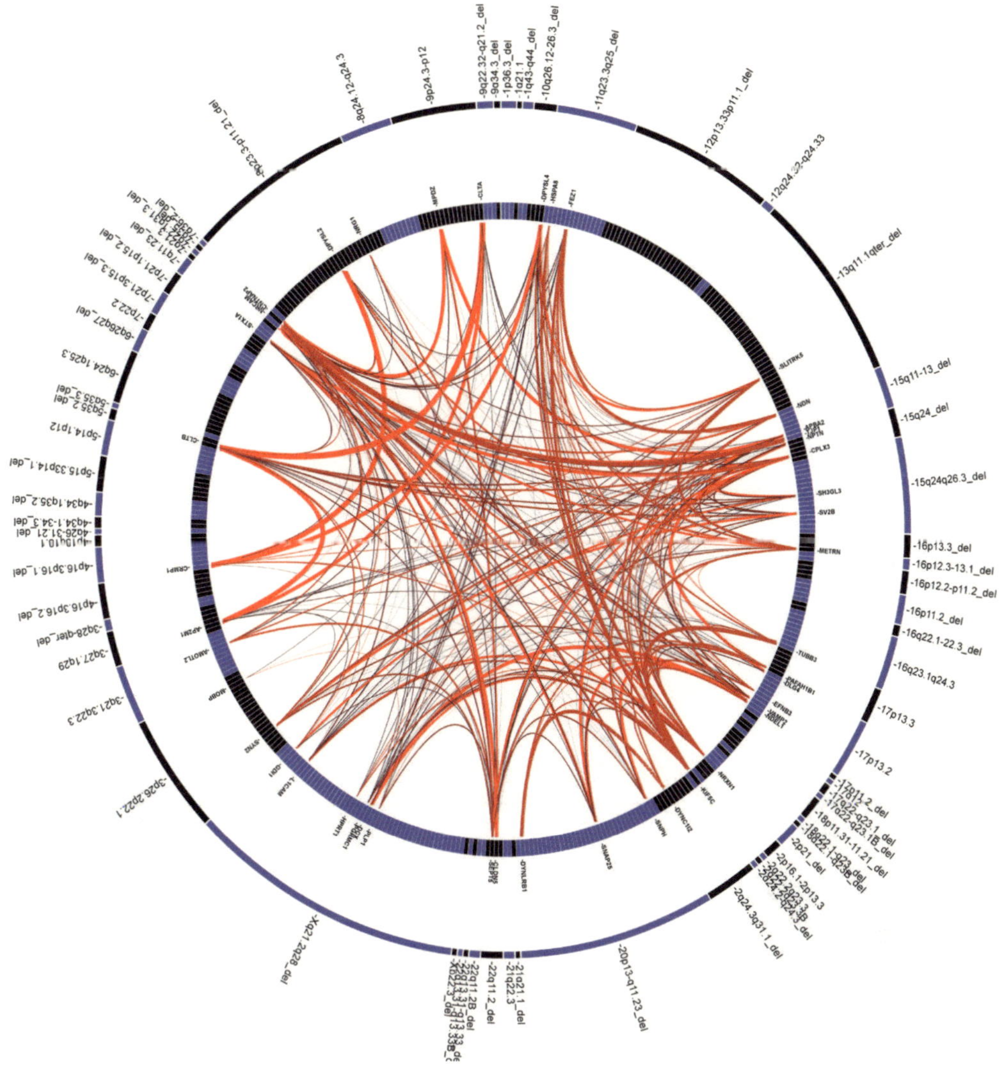

Figure 3. *in silico* bioinformatic mining of cross-loci gene interaction for intellectual disability. The double circle plots depict the gene relationships across the abnormal regions by GRAIL based on gene ontology data source. Outer circle represents abnormal regions, of which the G-band regions with '_del' are denoted deletions and the other regions are duplications. The inner circle stands for genes located in the corresponding regions in the outer circle, and for clear visualization only significant genes (p<0.001] and their connections are shown. The lines represent pair-wise gene connections, with the thickness of the lines indicating the degree functional similarity

In vivo animal models generated by genetic manipulation have also been used to study cytogenomic abnormalities. Mouse models of 16p11.2 deletion and duplication detected *in vivo* brain anomalies and behavior disorders [102]. Overexpression and transcript suppression of the 29 candidate genes from this 16p11.2 region in zebrafish identified the KCTD13 gene as a major driver of the mirrored neuroanatomical phenotypes [103]. Although the physiology and genetic makeup in the model animals are different from human, the animal models allow direct evaluation of gene-dosage effects and association of neuroanatomical defects with phenotypes. Figure 4 shows modeling of human cytogenomic abnormalities using stem cell technology and animal model and the potential application in drug development.

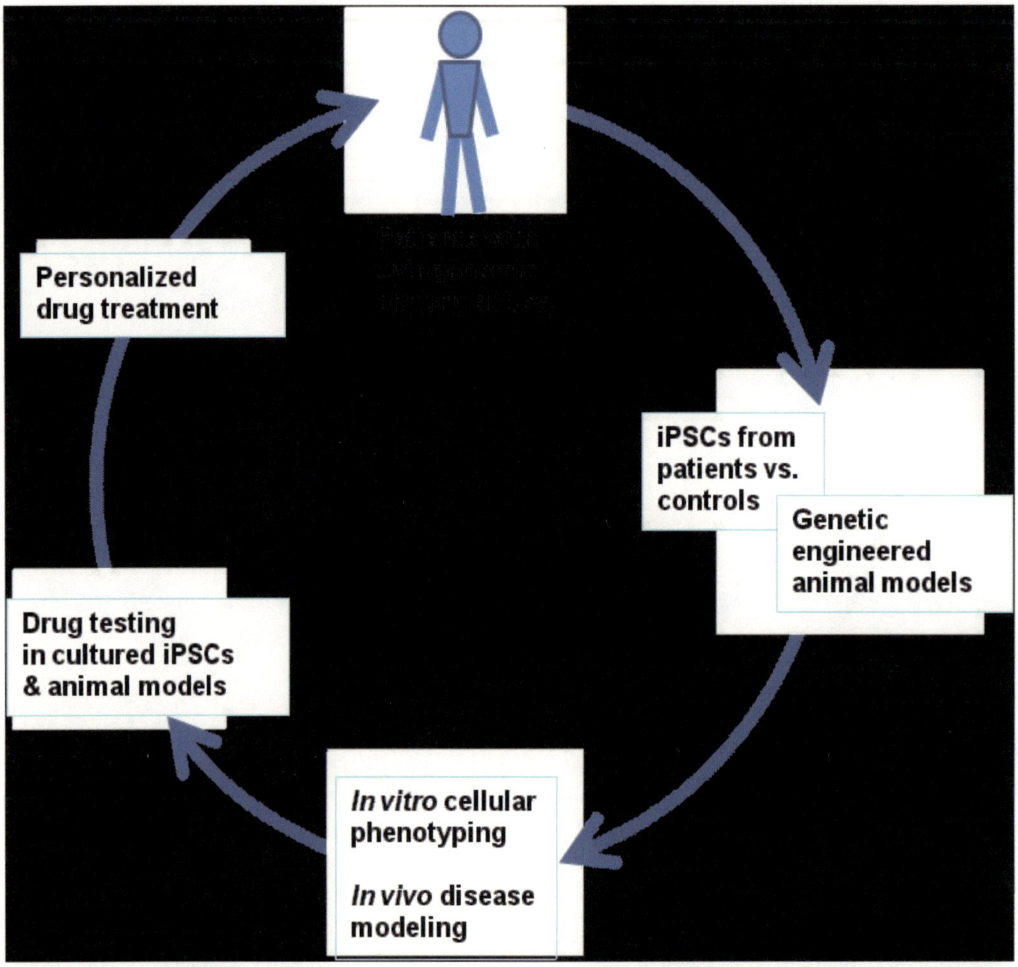

Figure 4. Using *in vitro* cellular phenotyping and *in vivo* animal modeling for cytogenomic abnormalities. To study gene functions and disease causing mechanims, iPSCs derived from affected patients with detected pCNVs are used for *in vitro* cellular phenotyping and genetic engineered animal models are used for *in vivo* disease pathophysiology. This cellular and animal models can also be used to screen drug for potential treatment of cytogenomic abnormalities.

5. Future directions and concluding remarks

In the first decade since the completion of the Human Genome Project, we have witnessed the rapid development of genomic technologies and the integration of genomic analysis into pediatric genetic evaluation. The experiences from current genomic analysis revealed a systematic approach including the evaluation of analytical validity of novel technologies using ROC statistics, the assessment of clinical validity through case series or clinical trials, the establishment of clinical evidence of detected genomic variants by large scale case-control studies, and the development of practice standards and guidelines as well as web-deliverable databases and resources. This approach will be effective for the integration of the next-generation exome sequencing and the next-next generation genomic sequencing technologies into clinical screening and diagnosis. However, the increased technical complexity of exome or genomic sequencing and the intellectual challenge to interpret a large amount of sequencing data will require more collaborative effort and supportive resources. The ultimate goal for an integrated pediatric and prenatal genetic evaluation is to provide comprehensive and in-depth profiling of karyotype, pCNV and mutations and associated medical phenotypes and risks for patients.

The aCGH or SNP chip analysis has brought pediatric genetic evaluation into the genomic era. This progress has contributed greatly to our understanding of genetic etiology in 12%-20% of pediatric patients with DD/ID/MCA/ASD. In additional to the technical progress in genetic diagnosis, the implementations of knowledge-based genetic counseling, rational clinical action and follow up familial studies could be of direct benefit for a substantial proportion of patients [104, 105]. For example, the aggressive behavior from patients with a 15q13.3 deletion involving the CHRNA7 gene could benefit from treatment with the NChR allosteric modulator and acetylcholinesterase (AChE) inhibitor, galantamine [106]. As we gain more knowledge of these genomic abnormalities through functional analysis using *in vitro* cellular and *in vivo* animal models, disease-specific guidance management and treatment could be developed, culminating in fully personalized medicine.

Web resources

Database of Genomic Variants (DGV): http://projects.tcag.ca/variation/

International Standards for Cytogenomic Arrays (ISCA): https://www.iscaconsortium.org/

DatabasE of Chromosomal Imbalance and Phenotype in Humans using Ensembl Resources (DECIPHER): http://decipher.sanger.ac.uk/

Human Genome Browser: http://genome.ucsc.edu/

Online Mendelian Inheritance in Man: http://www.ncbi.nlm.nih.gov/omim

Database for Annotation, Visualization, and Integrated Discovery (DAVID) (http://david.abcc.ncifcrf.gov/)

Gene Relationships Across Implicated Loci (http://www.broadinstitute.org/mpg/grail/)

Ingenuity Pathways Analysis (Ingenuity Systems Inc. http://www.ingenuity.com/)

Author details

Fang Xu and Peining Li*

Laboratory of Molecular Cytogenetics and Genomics, Department of Genetics, Yale University School of Medicine, New Haven, USA

References

[1] Smeets DFCM. Historical prospective of human cytogenetics: from microscope to microarray. Clinical Biochemistry 2004;37:439-46.

[2] Li MM, Andersson HC. Clinical application of microarray-based molecular cytogenetics: An emerging new era of genomic medicine. The Journal of Pediatrics 2009;155:311-7.

[3] Grayton HM, Fernandes C, Rujescu D, Collier DA. Copy number variations in neurodevelopmental disorders. Progress in Neurobiology 2012;99:81-91.

[4] Tjio JH, Levan A. The chromosome number of Man. Hereditas 1956;42:1-6.

[5] Hsu TC. Mammalian chromosome *in vitro*: I. The karyotype of Man. Journal of Heredity 1952;43:167-72.

[6] Lejeune J, Gautier M, Turpin R. Etude des chromosomes somatiques de neuf enfants mongoliens. Comptes Rendus 1959;248:17211-22.

[7] Ford CE, Jones KW, Polani PE, de Almeida JC, Briggs JH.. A sex-chromosome anomaly in a case of gonadal dysgenesis (Turner's syndrome). Lancet 1959; 1:711–3.

[8] Jacobs PA, Strong JA. A case of human intersexuality having a possible XXY sex-determining mechanism. Nature 1959;183:302–303.

[9] Patau K, Smith DW, Therman E, Inhorn SI, Wagner HP. Multiple congenital anomaly caused by an extra autosome. Lancet 1960;1:790–3.

[10] Edwards JH, Harnden DG, Cameron AH, Crosse VM, Wolff OH. A new trisomic syndrome. Lancet 1960; 1:787-790.

[11] Caspersson T, Farber S, Foley GE, Kudynowski J, Modest EJ, Simonsson E, Wagh U, Zech L. Chemical differentiation along metaphase chromosomes. Experimental Cell Research 1968;49:219–22.

[12] Shaffer LG, McGowan-Jordan J, Schmid M (eds). ISCN (2013): An international system for human cytogenetic nomenclature, S.Karger, Basel 2013.

[13] Wyandt HE, Tonk VS (eds). Atlas of human chromosome heteromorphisms. Kluwer Academic Publishers; 2004.

[14] Brothman AR, Schneider NR, Saikevych I, Cooley LD, Butler MG, Patil S, Mascarello JT, Rao KW, Dewald GW, Park JP, Persons DL, Wolff DJ, Vance GH. Cytogenetics Resource Committee, College of American Pathologists/American College of Medical

Genetics. Cytogenetic heteromorphisms: survey results and reporting practices of giemsa-band regions that we have pondered for years. Archives Pathology & Laboratory Medicine 2006; 130:947-9.

[15] Langer-Safer PR, Levine M, Ward DC. Immunological method for mapping genes on Drosophila polytene chromosomes. Proceedings of the National Academy of Sciences USA 1982;79(14):4381-5.

[16] Ried T, Landes G, Dackowski W, Klinger K, Ward DC. Multicolor fluorescence in situ hybridization for the simultaneous detection of probe sets for chromosomes 13, 18, 21, X and Y in uncultured amniotic fluid cells. Human Molecular Genetics 1992;5:307-13.

[17] Ning Y, Roschke A, Smith AC, Macha M, Precht K, Riethman H, Ledbetter DH, Flint J, Horsley S, Regan R, Kearney L, Knight S, Kvaloy K, Brown WRA. A complete set of human telomeric probes and their clinical application. Nature Genetics 1996; 14:86-9.

[18] Shaffer LG, American College of Medical Genetics Professional Practice and Guidelines Committee. American College of Medical Genetics guideline on the cytogenetic evaluation of the individual with developmental delay or mental retardation. Genetics in Medicine 2005; 7:650-4.

[19] Kallioniemi A, Kallioniemi OP, Sudar D, Rutovitz D, Gray JW, Waldman F, Pinkel D. Comparative genomic hybridization for molecular cytogenetic analysis of solid tumors. Science 1992; 258:818-21.

[20] Schena M, Shalon D, Davis RW, Brown PO. Quantitative monitoring of gene expression patterns with a complementary DNA microarray. Science 1995; 270:467–70.

[21] Pinkel D, Segraves R, Sudar D. High resolution analysis of DNA copy number variation using comparative genomic hybridization to microarrays. Nature Genetics 1998; 20:207–11.

[22] Snijders AM, Nowak N, Segraves R, Blackwood S, Brown N, Conroy J, Hamilton G, Hindle AK, Huey B, Kimura K, Law S, Myambo K, Palmer J, Ylstra B, Yue JP, Gray JW, Jain AN, Pinkel D, Albertson DG. 2001. Assembly of microarrays for genome-wide measurement of DNA copy number. Nature Genetics 2001; 29:263–64.

[23] Iafrate AJ, Feuk L, Rivera MN, Listwnik ML, Donahoe PK, Qi Y, Scherer SW, Lee C. Detection of large-scale variation in the human genome. Nature Genetics 2004; 36:949-51.

[24] Redon R, Ishikawa S, Fitch KR, Feuk L, Perry GH, Andrews TD, Fiegler H, Shapero MH, Carson AR, Chen W, Cho EK, Dallaire S, Freeman JL, González JR, Gratacòs M, Huang J, Kalaitzopoulos D, Komura D, MacDonald JR, Marshall CR, Mei R, Montgomery L, Nishimura K, Okamura K, Shen F, Somerville MJ, Tchinda J, Valsesia A, Woodwark C, Yang F, Zhang J, Zerjal T, Zhang J, Armengol L, Conrad DF, Estivill X, Tyler-Smith C, Carter NP, Aburatani H, Lee C, Jones KW, Scherer SW, Hurles ME. Global variation in copy number in the human genome. Nature 2006; 444:444-54.

[25] Vissers LE, de Vries BB, Osoegawa K, Janssen IM, Feuth T, Choy CO, Straatman H, van der Vliet W, Huys EH, van Rijk A, Smeets D, van Ravenswaaij-Arts CM, Knoers

NV, van der Burgt I, de Jong PJ, Brunner HG, van Kessel AG, Schoenmakers EF, Veltman JA. Array-based comparative genomic hybridization for the genomewide detection of submicroscopic chromosomal abnormalities. American Journal of Human Genetics 2003; 73:1261-70.

[26] Shaw-Smith C, Redon R, Rickman L, Rio M, Willatt L, Fiegler H, Firth H, Sanlaville D, Winter R, Colleaux L, Bobrow M, Carter NP. Microarray based comparative genomic hybridisation (array-CGH) detects submicroscopic chromosomal deletions and duplications in patients with learning disability/mental retardation and dysmorphic features. Journal of Medical Genetics 2004; 41:241–8.

[27] Schoumans J, Ruivenkamp C, Holmberg E, Kyllerman M, Anderlid BM, Nordenskjold M. Detection of chromosomal imbalances in children with idiopathic mental retardation by array based comparative genomic hybridization (array-CGH). Journal of Medical Genetics 2005; 42:699-705.

[28] Friedman JM, Baross A, Delaney AD, Ally A, Arbour L, Armstrong L, Asano J, Bailey DK, Barber S, Birch P, Brown-John M, Cao M, Chan S, Charest DL, Farnoud N, Fernandes N, Flibotte S, Go A, Gibson WT, Holt RA, Jones SJ, Kennedy GC, Krzywinski M, Langlois S, Li HI, McGillivray BC, Nayar T, Pugh TJ, Rajcan-Separovic E, Schein JE, Schnerch A, Siddiqui A, Van Allen MI, Wilson G, Yong SL, Zahir F, Eydoux P, Marra MA. Oligonucleotide microarray analysis of genomic imbalances in children with mental retardation. American Journal of Human Genetics 2006; 79:500-13.

[29] Holtzman NA, Watson MS. Promoting safe and effective genetic testing in the United States: Final report of the task force on genetic testing. The Johns Hopkins University Press, 1998.

[30] Yu W, Ballif BC, Kashork CD, Heilstedt HA, Howard LA, Cai WW, White LD, Liu W, Beaudet AL, Bejjani BA, Shaw CA, Shaffer LG. Development of a comparative genomic hybridization microarray and demonstration of its utility with 25 well-characterized 1p36 deletions. Human Molecular Genetics 2003; 12:2145-52.

[31] Cheung SW, Shaw CA, Yu W, Li J, Ou Z, Patel A, Yatsenko SA, Cooper ML, Furman P, Stankiewicz P, Lupski JR, Chinault AC, Beaudet AL. Development and validation of aCGH microarray for clinical cytogenetic diagnosis. Genetics in Medicine 2005; 7: 422-32.

[32] Xiang B, Li A, Valentin D, Novak N, Zhao H-Y, Li P. 2008. Analytical and clinical validity of whole genome oligonucleotide array comparative genomic hybridization for pediatric patients with mental retardation and developmental delay. American Journal of Medical Genetics 2008; 146A: 1942-54.

[33] Baldwin EL, Lee JY, Blake DM, Bunke BP, Alexander CR, Kogan AL, Ledbetter DH, Martin CL. Enhanced detection of clinically relevant genomic imbalances using a targeted plus whole genome oligonucleotide microarray. Genetics in Medicine 2008;10:415-29.

[34] Neill NJ, Torchia BS, Bejjani BA, Shaffer LG, Ballif BC. Comparative analysis of copy number detection by whole-genome BAC and oligonucleotide array CGH. Molecular Cytogenetics 2010;3:11.

[35] Zhang ZF, Ruivenkamp C, Staaf J, Zhu H, Barbaro M, Petillo D, Khoo SK, Borg A, Fan YS, Schoumans J. Detection of submicroscopic constitutional chromosome aberrations in clinical diagnostics: a validation of the practical performance of different array platforms. European Journal of Human Genetics 2008;16:786-92.

[36] Schaaf CP, Wiszniewska J, Beaudet AL. Copy number and SNP arrays in clinical diagnostics. Annual Review of Genomics and Human Genetics 2011;12:25-51.

[37] Papenhausen P, Schwartz S, Risheg H, Keitges E, Gadi I, Burnside RD, Jaswaney V, Pappas J, Pasion R, Friedman K, Tepperberg J, UPD detection using homozygosity profiling with a SNP genotyping microarray. American Journal of Medical Genetics 2011;155A:757-68.

[38] Tucker T, Schlade-Bartusiak K, Eydoux P, Nelson TN, Brown L. Uniparental disomy: can SNP array data be used for diagnosis? Genetics in Medicine 2012;14:753-56.

[39] Boone PM, Bacino CA, Shaw CA, Eng PA, Hixson PM, Pursley AN, Kang SH, Yang Y, Wiszniewska J, Nowakowska BA, del Gaudio D, Xia Z, Simpson-Patel G, Immken LL, Gibson JB, Tsai AC, Bowers JA, Reimschisel TE, Schaaf CP, Potocki L, Scaglia F, Gambin T, Sykulski M, Bartnik M, Derwinska K, Wisniowiecka-Kowalnik B, Lalani SR, Probst FJ, Bi W, Beaudet AL, Patel A, Lupski JR, Cheung SW, Stankiewicz P. Detection of clinically relevant exonic copy-number changes by array CGH. Human Mutation 2010;31:1326-42.

[40] Aradhya S, Lewis R, Bonaga T, Nwokekeh N, Stafford A, Boggs B, Hruska K, Smaoui N, Compton JG, Richard G, Suchy S. Exon-level array CGH in a large clinical cohort demonstrates increased sensitivity of diagnostic testing for Mendelian disorders. Genetics in Medicine 2012;14:594-603.

[41] Shearer BM, Thorland EC, Gonzales PR, Ketterling RP. Evaluation of a commercially available focused aCGH platform for the detection of constitutional chromosome anomalies. American Journal of Medical Genetics 2007;143A:2357-70.

[42] Shaffer LG, Beaudet AL, Brothman AR, Hirsch B, Levy B, Martin CL, Mascarello JT, Rao KW, Working Group of the Laboratory Quality Assurance Committee of the American College of Medical Genetics. Microarray analysis for constitutional cytogenetic abnormalities. Genetics in Medicine 2007;9:654-62.

[43] Shen Y, Irons M, Miller DT, Cheung SW, Lip V, Sheng X, Tomaszewicz K, Shao H, Fang H, Tang HS, Irons M, Walsh CA, Platt O, Gusella JF, Wu BL. Development of a focused oligonucleotide-array comparative genomic hybridization chip for clinical diagnosis of genomic imbalance. Clinical Chemistry 2007;53:2051-9.

[44] Yu S, Bittel DC, Kibiryeva N, Zwick DL, Cooley LD. Validation of the Agilent 244K oligonucleotide array-based comparative genomic hybridization platform for clinical cytogenetic diagnosis. American Journal of Clinical Pathology 2009;132:349-60.

[45] Xiang B, Zhu H, Shen Y, Miller DT, Lu K, Hu X, Andersson HC, Narumanchi TM, Wang Y, Martinez JE, Wu B-L, Li P, Li MM, Chen T-J, Fan Y-S. Genome-wide oligonucleotide array CGH for etiological diagnosis of mental retardation: A multi-center experience of 1,499 clinical cases. Journal of Molecular Diagnosis 2010;12:204-12.

[46] Kaminsky EB, Kaul V, Paschall J, Church DM, Bunke B, Kunig D, Moreno-De-Luca D, Moreno-De-Luca A, Mulle JG, Warren ST, Richard G, Compton JG, Fuller AE, Gliem TJ, Huang S, Collinson MN, Beal SJ, Ackley T, Pickering DL, Golden DM, Aston E, Whitby H, Shetty S, Rossi MR, Rudd MK, South ST, Brothman AR, Sanger WG, Iyer RK, Crolla JA, Thorland EC, Aradhya S, Ledbetter DH, Martin CL. An evidence-based approach to establish the functional and clinical significance of copy number variants in intellectual and developmental disabilities. Genetics in Medicine 2011;13:777-84.

[47] Miller DT, Adam MP, Aradhya S, Biesecker LG, Brothman AR, Carter NP, Church DM, Crolla JA, Eichler EE, Epstein CJ, Faucett WA, Feuk L, Friedman JM, Hamosh A, Jackson L, Kaminsky EB, Kok K, Krantz ID, Kuhn RM, Lee C, Ostell JM, Rosenberg C, Scherer SW, Spinner NB, Stavropoulos DJ, Tepperberg JH, Thorland EC, Vermeesch JR, Waggoner DJ, Watson MS, Martin CL, Ledbetter DH. Consensus statement: chromosomal microarray is a first-tier clinical diagnostic test for individuals with developmental disabilities or congenital anomalies. American Journal of Human Genetics 2010;86:749-64.

[48] Manning M, Hudgins L, Professional Practice and Guidelines Committee. Array-based technology and recommendations for utilization in medical genetics practice for detection of chromosomal abnormalities. Genetics in Medicine 2010;12:742-5.

[49] Kearney HM, Thorland EC, Brown KK, Quintero-Rivera F, South ST, Working Group of the American College of Medical Genetics Laboratory Quality Assurance Committee. American College of Medical Genetics standards and guidelines for interpretation and reporting of postnatal constitutional copy number variants. Genetics in Medicine 2011;13:680-5.

[50] Kearney HM, South ST, Wolff DJ, Lamb A, Hamosh A, Rao KW, Working Group of the American College of Medical Genetics. American College of Medical Genetics recommendations for the design and performance expectations for clinical genomic copy number microarrays intended for use in the postnatal setting for detection of constitutional abnormalities. Genetics in Medicine 2011;13:676-9.

[51] Lee C, Iafrate AJ, Brothman AR. Copy number variations and clinical cytogenetic diagnosis of constitutional disorders. Nature Genetics 2007;39:S48-54.

[52] Tsuchiya KD, Shaffer LG, Aradhya S, Gastier-Foster JM, Patel A, Rudd MK, Biggerstaff JS, Sanger WG, Schwartz S, Tepperberg JH, Thorland EC, Torchia BA, Brothman AR. Variability in interpreting and reporting copy number changes detected by array-based technology in clinical laboratories. Genetics in Medicine 2009;11:866-73.

[53] de Leeuw N, Dijkhuizen T, Hehir-Kwa JY, Carter NP, Feuk L, Firth HV, Kuhn RM, Ledbetter DH, Martin CL, van Ravenswaaij-Arts CM, Scherer SW, Shams S, Van Vooren S, Sijmons R, Swertz M, Hastings R. Diagnostic interpretation of array data using public databases and internet sources. Human Mutation 2012;33:930-40.

[54] Corpas M, Bragin E, Clayton S, Bevan P, Firth HV. Interpretation of genomic copy number variants using DECIPHER. In: Current Protocol of Human Genetics, Chapter 8:Unit 8.14,2012.

[55] Riggs ER, Church DM, Hanson K, Horner VL, Kaminsky EB, Kuhn RM, Wain KE, Williams ES, Aradhya S, Kearney HM, Ledbetter DH, South ST, Thorland EC, Martin CL. Towards an evidence-based process for the clinical interpretation of copy number variation. Clinical Genetics 2011;19:1-10.

[56] Paciorkowski AR, Fang M. Chromosomal microarray interpretation: what is a child neurologist to do? Pediatric Neurology 2009;41:391-8.

[57] Brothman AR, Dolan MM, Goodman BK, Park JP, Persons DL, Saxe DF, Tepperberg JH, Tsuchiya KD, Van Dyke DL, Wilson KS, Wolff DJ, Theil KS. College of American Pathologists/American College of Medical Genetics proficiency testing for constitutional cytogenomic microarray analysis. Genetics in Medicine 2011;13, 765-9.

[58] Shevell M, Ashwal S, Donley D, Flint J, Gingold M, Hirtz D, Donley D, Flint J, Gingold M, Hirtz D, Majnemer A, Noetzel M, Sheth RD; Quality Standards Subcommittee of the American Academy of Neurology; Practice Committee of the Child Neurology Society. Parctice parameter: evaluation of the child with global developmental delay: report of the quality standards subcommittee of the child neurology society. Neurology 2003;60:367-80.

[59] Ledbetter DH, Martin CL. Cryptic telomere imbalance: a 15-year update. American Journal of Medical Genetics 2007;145C: 327-34.

[60] Rossi MR, DiMaio M, Xiang B, Lu K, Hande K, Seashore G, Mahoney M, Li P. Clinical and genomic characterization of distal duplications and deletions of chromosome 4q: Study of two cases and review of the literature. American Journal of Medical Genetics 2009;149A:2788-94.

[61] Khattab M, Xu F, Li P, Bhandari V. A de novo 3.54 Mb deletion of 17q22-q23.1 associated with hydrocephalus: A case report and review of literature. American Journal of Medical Genetics 2011;155A:3082-6.

[62] van Bon BW, Koolen DA, Borgatti R, Magee A, Garcia-Minaur S, Rooms L, Reardon W, Zollino M, Bonaglia MC, De Gregori M, Novara F, Grasso R, Ciccone R, van Duyvenvoorde HA, Aalbers AM, Guerrini R, Fazzi E, Nillesen WM, McCullough S, Kant SG, Marcelis CL, Pfundt R, de Leeuw N, Smeets D, Sistermans EA, Wit JM, Hamel BC, Brunner HG, Kooy F, Zuffardi O, de Vries BB. Clinical and molecular characteristics of 1qter microdeletion syndrome: delineating a critical region for corpus callosum agenesis/hypogenesis. Journal of Medical Genetics 2008;45(6):346-54.

[63] Orellana C, Roselló M, Monfort S, Oltra S, Quiroga R, Ferrer I, Martínez F. Corpus callosum abnormalities and the controversy about the candidate genes located in 1q44. Cytogenetics and Genome Research 2009;127:5-8.

[64] Caliebe A, Kroes HY, van der Smagt JJ, Martin-Subero JI, Tönnies H, van 't Slot R, Nievelstein RA, Muhle H, Stephani U, Alfke K, Stefanova I, Hellenbroich Y, Gillessen-Kaesbach G, Hochstenbach R, Siebert R, Poot M. Four patients with speech delay, seizures and variable corpus callosum thickness sharing a 0.440 Mb deletion in region 1q44 containing the HNRPU gene. European Journal of Medical Genetics 2010;53:179-85.

[65] Thierry G, Bénéteau C, Pichon O, Flori E, Isidor B, Popelard F, Delrue MA, Duboscq-Bidot L, Thuresson AC, van Bon BW, Cailley D, Rooryck C, Paubel A, Metay C, Dusser A, Pasquier L, Béri M, Bonnet C, Jaillard S, Dubourg C, Tou B, Quéré MP, Soussi-Zander C, Toutain A, Lacombe D, Arveiler B, de Vries BB, Jonveaux P, David A, Le Caignec C. Molecular characterization of 1q44 microdeletion in 11 patients reveals three candidate genes for intellectual disability and seizures. American Journal of Medical Genetics 2012;158A:1633-40.

[66] Shimojima K, Okamoto N, Suzuki Y, Saito M, Mori M, Yamagata T, Momoi MY, Hattori H, Okano Y, Hisata K, Okumura A, Yamamoto T. Subtelomeric deletions of 1q43q44 and severe brain impairment associated with delayed myelination. Journal of Human Genetics 2012;57:593-600

[67] Ballif, B. C., Rosenfeld, J. A., Traylor, R., Theisen, A., Bader, P. I., Ladda, R. L., Sell, S. L., Steinraths, M., Surti, U., McGuire, M., Williams, S., Farrell, S. A., and 18 others. High-resolution array CGH defines critical regions and candidate genes for microcephaly, abnormalities of the corpus callosum, and seizure phenotypes in patients with microdeletions of 1q43q44. Human Genetics 2012;131:145-56.

[68] Selmer KK, Bryne E, Rødningen OK, Fannemel M. A de novo 163 kb interstitial 1q44 microdeletion in a boy with thin corpus callosum, psychomotor delay and seizures. European Journal of Medical Genetics 2012;55:715-8.

[69] Guilherme RS, Meloni VF, Kim CA, Pellegrino R, Takeno SS, Spinner NB, Conlin LK, Christofolini DM, Kulikowski LD, Melaragno MI. Mechanisms of ring chromosome formation, ring instability and clinical consequences. BMC Medical Genetics 2011;12:171.

[70] Zhang HZ, Xu F, Seashore M, Li P. Unique genomic structure and distinct mitotic behavior of ring chromosome 21 in two unrelated cases. Cytogenetics and Genome Research. 2012;136:180-7.

[71] Yu S, Fiedler SD, Brawner SJ, Joyce JM, Zhou XG, Liu HY. Characterizing small supernumerary marker chromosomes with combination of multiple techniques. Cytogenetics and Genome Research 2012;136:6-14.

[72] Conlin LK, Kramer W, Hutchinson AL, Li X, Riethman H, Hakonarson H, Mulley JC, Scheffer IE, Berkovic SF, Hosain SA, Spinner NB. Molecular analysis of ring chromosome 20 syndrome reveals two distinct groups of patients. Journal of Medical Genetics. 2011;48:1-9.

[73] Ballif BC, Rorem EA, Sundin K, Lincicum M, Gaskin S, Coppinger J, Kashork CD, Shaffer LG, Bejjani BA. Detection of low-level mosaicism by array CGH in routine diagnostic specimens. American Journal of Medical Genetics 2006;140A:2757-67.

[74] Cheung SW, Shaw CA, Scott DA, Patel A, Sahoo T, BAcino CA, Pursley A, Li J, Erickson R, Gropman L, Miller DT, Seashore MR, Summers AM, Stankiewicz P, Chinault AC, Lupski JR, Beaudet AL, Sutton VR. Microarray-based CGH detects chromosomal mosaicism not revealed by conventional cytogenetics. American Journal of Medical Genetics 2007;143A:1679-86.

[75] Li P, Zhang HZ, Huff S, Nimmakayalu M, Qumsiyeh M, Yu JW, Szekely A, Xu T, Pober BR. Karyotype-phenotype insights from 11q14.1-q23.2 interstitial deletions: *FZD4* haploinsufficiency and exudative vitroretinopathy in a patient with a complex chromosome rearrangement. American Journal of Medical Genetics 2006;140A: 2721-9.

[76] Higgins AW, Alkuraya FS, Bosco AF, Brown KK, Bruns GA, Donovan DJ, Eisenman R, Fan Y, Farra CG, Ferguson HL, Gusella JF, Harris DJ, Herrick SR, Kelly C, Kim HG, Kishikawa S, Korf BR, Kulkarni S, Lally E, Leach NT, Lemyre E, Lewis J, Ligon AH, Lu W, Maas RL, MacDonald ME, Moore SD, Peters RE, Quade BJ, Quintero-Rivera F, Saadi I, Shen Y, Shendure J, Williamson RE, Morton CC. Characterization of apparently balanced chromosomal rearrangements from the developmental genome anatomy project. American Journal Human Genetics 2008;82:712-22.

[77] De Gregori M, Ciccone R, Magini P, Pramparo T, Gimelli S, Messa J, Novara F, Vetro A, Rossi E, Maraschio P, Bonaglia MC, Anichini C, Ferrero GB, Silengo M, Fazzi E, Zatterale A, Fischetto R, Previderé C, Belli S, Turci A, Calabrese G, Bernardi F, Meneghelli E, Riegel M, Rocchi M, Guerneri S, Lalatta F, Zelante L, Romano C, Fichera M, Mattina T, Arrigo G, Zollino M, Giglio S, Lonardo F, Bonfante A, Ferlini A, Cifuentes F, Van Esch H, Backx L, Schinzel A, Vermeesch JR, Zuffardi O. 2007. Cryptic deletions are a common finding in "balanced" reciprocal and complex chromosome rearrangements: A study of 59 patients. Journal of Medical Genetics 2007;44:750–62.

[78] Baptista J, Mercer C, Prigmore E, Gribble SM, Carter NP, Maloney V, Thomas NS, Jacobs PA, Crolla JA, Breakpoint mapping and array CGH in translocations: comparison of a phenotypically normal and an abnormal cohort. American Journal of Human Genetics 2008;82:927–936.

[79] Cacciagli P, Haddad MR, Mignon-Ravix C, El-Waly B, Moncla A, Missirian C, Chabrol B, Villard L. Disruption of the ATP8A2 gene in a patient with a t(10;13) de novo balanced translocation and a severe neurological phenotype. European Journal of Human Genetics 2010;18:1360-3.

[80] Brownstein CA, Adler F, Nelson-Williams C, Lijima J, Li P, Imura A, Nabeshima Y, Reyes-Mugica M, Carpenter TO, Lifton RP. A translocation causing increased α-Klotho level results in hypophosphatemic rickets and hyperparathyroidism. Proceedings of the National Academy of Sciences USA 2008;105:3455-60.

[81] Talkowski ME, Ernst C, Heilbut A, Chiang C, Hanscom C, Lindgren A, Kirby A, Liu S, Muddukrishna B, Ohsumi TK, Shen Y, Borowsky M, Daly MJ, Morton CC, Gusella JF. Next-generation sequencing strategies enable routine detection of balanced chromosome rearrangements for clinical diagnostics and genetic research. American Journal of Human Genetics 2011;88:469-81.

[82] Koolen DA, Vissers LE, Pfundt R, de Leeuw N, Knight SJ, Regan R, Kooy RF, Reyniers E, Romano C, Fichera M, Schinzel A, Baumer A, Anderlid BM, Schoumans J, Knoers NV, van Kessel AG, Sistermans EA, Veltman JA, Brunner HG, de Vries BB. A new chromosome 17q21.31 microdeletion syndrome associated with a common inversion polymorphism. Nature Genetics 2006;38:999-1001.

[83] Shaw-Smith C, Pittman AM, Willatt L, Martin H, Rickman L, Gribble S, Curley R, Cumming S, Dunn C, Kalaitzopoulos D, Porter K, Prigmore E, Krepischi-Santos AC, Varela MC, Koiffmann CP, Lees AJ, Rosenberg C, Firth HV, de Silva R, Carter NP. Microdeletion encompassing MAPT at chromosome 17q21.3 is associated with developmental delay and learning disability. Nature Genetics 2006;38:1032-37.

[84] Kirchhoff M. Bisgaard A-M, Duno M, Hansen FJ, Schwartz MA. 17q21.31 microduplication, reciprocal to the newly described 17q21.31 microdeletion, in a girl with severe psychomotor developmental delay and dysmorphic craniofacial features. European Journal of Medical Genetics 2007;50:256-63.

[85] Cooper GM, Coe BP, Girirajan S, Rosenfeld JA, Vu TH, Baker C, Williams C, Stalker H, Hamid R, Hannig V, Abdel-Hamid H, Bader P, McCracken E, Niyazov D, Leppig K, Thiese H, Hummel M, Alexander N, Gorski J, Kussmann J, Shashi V, Johnson K, Rehder C, Ballif BC, Shaffer LG, Eichler EE. A copy number variation morbidity map of developmental delay. Nature Genetics 2011;43:838-46.

[86] Donnelly MP, Paschou P, Grigorenko E, Gurwitz D, Mehdi SQ, Kajuna SLB, Barta C, Kungulilo S, Karoma NJ, Lu R-B, Zhukova OV, Kim J-J, Comas D, Sinicalco M, New M, Li P, Li H, Speed WC, Rajeevan H, Pakstis A, Kidd JR, Kidd KK. The distribution and most recent common ancestor of the 17q21 inversion in humans. American Journal of Human Genetics 2010;86:161-171.

[87] Merla G, Brunetti-Pierri N, Micale, Fusco C. Copy number variants at Williams-Beuren syndrome 7q11.23 regions. Hum Genet 2010;128:3-26.

[88] Weiss LA, Shen Y, Korn JM, Arking DE, Miller DT, Fossdal R, Saemundsen E, Stefansson H, Ferreira MA, Green T, Platt OS, Ruderfer DM, Walsh CA, Altshuler D, Chakravarti A, Tanzi RE, Stefansson K, Santangelo SL, Gusella JF, Sklar P, Wu BL, Daly MJ; Autism Consortium. Association between microdeletion and microduplication at 16p11.2 and autism. New England Journal of Medicine 2008;358(7):667-75.

[89] Jacquemont S, Reymond A, Zufferey F, Harewood L, Walters RG, Kutalik Z, Martinet D et al. Mirror extreme BMI phenotypes associated with gene dosage at the chromosome 16p11.2 locus. Nature 2011;478:97-102.

[90] Mefford H, Sharp A, Baker C, Itsara A, Jiang Z, Buysse K, Huang S, Maloney V, Crolla J, Baralle D, Collins A, Mercer C, et al. Recurrent rearrangements of chromosome 1q21.1 and variable pediatric phenotypes. New England Journal of Medicine 2008;359:1685-99.

[91] Brunetti-Pierri N, Berg JS, Scaglia F, Belmont J, Bacino CA, Sahoo T, Lalani SR, Graham B, Lee B, Shinawi M, Shen J, Kang SH, Pursley A, Lotze T, Kennedy G, Lansky-Shafer S, Weaver C, Roeder ER, Grebe TA, Arnold GL, Hutchison T, Reimschisel T, Amato S, Geragthy MT, Innis JW, Obersztyn E, Nowakowska B, Rosengren SS, Bader PI, Grange DK, Naqvi S, Garnica AD, Bernes SM, Fong CT, Summers A, Walters WD, Lupski JR, Stankiewicz P, Cheung SW, Patel A. Recurrent reciprocal 1q21.1 deletions and duplications associated with microcephaly or macrocephaly and developmental and behavioral abnormalities. Nature Genetics 2008;40:1466-71.

[92] Sahoo T, Theisen A, Rosenfeld JA, Lamb A N, Ravnan JB, Schultz RA, Torchia BS, Neill N, Casci I, Bejjani BA, Shaffer LG. Copy number variants of schizophrenia sus-

ceptibility loci are associated with a spectrum of speech and developmental delays and behavior problems. Genetics in Medicine 2011;13:868-80.

[93] Carmona-Mora P, Walz K. Retinoic Acid Induced 1, RAI1: A Dosage Sensitive Gene Related to Neurobehavioral Alterations Including Autistic Behavior. Current Genomics 2010;11:607-17.

[94] Weksberg R, Shuman C, Beckwith JB. 2010. Beckwith-Wiedemann syndrome. European Journal of Human Genetics 2010;18:8–14.

[95] Qiao Y, Harvard C, Tyson C, Liu X, Fawcett C, Pavlidis P, Holden JJ, Lewis ME, Rajcan-Separovic E. Outcome of array CGH analysis for 255 subjects with intellectual disability and search for candidate genes using bioinformatics. Human Genetics 2010;128:179-94.

[96] Pinto D, Pagnamenta AT, Klei L, Anney R, Merico D, Regan R, Conroy J, et al. Functional impact of global rare copy number variation in autism spectrum disorders. Nature 2010;466:368-72.

[97] Xu F, Li L, Schulz VP, Gallagher PG, Xiang B, Zhao H-Y, Li P. Identification of candidate brain expressed genes and interacting networks for intellectual disability by integrated cytogenetic, genomic and bioinformatic analyses. The Global Conference of Chinese Geneticists. Hangzhou, China, 2012.

[98] Marchetto MC, Carromeu C, Acab A, Yu D, Yeo GW, Mu Y, Chen G, Gage FH, Muotri AR. A model for neural development and treatment of Rett syndrome using human induced pluripotent stem cells. Cell 2010;143:527-39.

[99] Yang J, Cai J, Zhang Y, Wang X, Li W, Xu J, Li F, Guo X, Deng K, Zhong M Chen Y, Lai L, Pei D, Esteban M. Induced pluripotent stem cells can be used to model the genomic imprinting disorder prader-willi syndrome. Journal Biological Chemistry 2010; 285:40303-11.

[100] Li W, Wang X, Fan W, Zhao P, Chan YC, Chen S, Zhang S, Guo X, Zhang Y, Li Y, Cai J, Qin D, Li X, Yang J, Peng T, Zychlinski D, Hoffmann D, Zhang R, Deng K, Ng KM, Menten B, Zhong M, Wu J, Li Z, Chen Y, Schambach A, Tse HF, Pei D, Esteban MA. Modeling abnormal early development with induced pluripotent stem cells from aneuploid syndromes. Human Molecular Genetics. 2012;21:32-45.

[101] Chailangkarn T, Acab A, Muotri AR. Modeling neurodevelopmental disorders using human neurons. Current Opinion in Neurobiology. 2012;22:1-6.

[102] Horev G, Ellegood J, Lerch JP, Son YE, Muthuswamy L, Vogel H, Krieger AM, Buja A, Henkelman RM, Wigler M, Mills AA. Dosage-dependent phenotypes in models of 16p11.2 lesions found in autism. Proceedings of the National Academy of Sciences USA 2011;108:17076-81.

[103] Golzio C, Willer J, Talkowski ME, Oh EC, Taniguchi Y, Jacquemont S, Reymond A, Sun M, Sawa A, Gusella JF, Kamiya A, Beckmann JS, Katsanis N. KCTD13 is a major driver of mirrored neuroanatomical phenotypes of the 16p11.2 copy number variant. Nature 2012;485:363-7.

[104] Darilek S, Ward P, Pursley A, Plunkett K, Furman P, Magoulas P, Patel A, Cheung SW, Eng CM. Pre- and postnatal genetic testing by array-comparative genomic hybridization: genetic counseling perspectives. Genetics in Medicine 2008;10:13-18.

[105] Coulter ME, Miller DT, Harris DJ, Hawley P, Picker J, Roberts AE, Sobeih MM, Irons M. Chromosomal microarray testing influences medical management. Genetics in Medicine 2011;13:770-6.

[106] Cubells JF, Deoreo EH, Harvey PD, Garlow SJ, Garber K, Adam MP, Martin CL. Pharmaco-genetically guided treatment of recurrent rage outbursts in an adult male with 15q13.3 deletion syndrome. American Journal of Medical Genetics 2011;155A: 805-10.

Chapter 2

Chromatin Structure and Intellectual Disability Syndromes

Adrienne Elbert and Nathalie G. Bérubé

Additional information is available at the end of the chapter

http://dx.doi.org/10.5772/55730

1. Introduction

The molecular complex consisting of DNA and its associated proteins is referred to as chromatin. In the central nervous system (CNS), dynamic chromatin remodelling is required for cell division, specification, differentiation, maturation and to respond appropriately to environmental cues (reviewed in 1-4). Modifications to chromatin can act as a form of cellular memory, storing information about a cell's development, differentiation and environment [5].

In humans, the cerebral cortex is required for normal memory, information processing, thought, attention, perception and language. It consists of six horizontal layers of excitatory pyramidal neurons interspersed with inhibitory interneurons that form distinct synaptic circuits. During synaptic transmission, neurotransmitters released by a neuron bind receptors and initiate electrical signals that travel through the axon of the neighbouring neuron, and in this process alter its morphology and behaviour. One of these modifiable neuronal behaviours is the strength of the synaptic response, termed synaptic plasticity. The complex morphological and gene expression changes that are triggered by synapse formation must be maintained so that the maturing neuron can develop its identity and specific role in the nervous system. Dynamic changes in chromatin structure and gene expression underlie many of the above processes. Perhaps not surprisingly, many neurodevelopmental syndromes characterized by intellectual disabilities are caused by mutations in chromatin modifying factors. In this chapter, we provide an overview of the basic concepts behind chromatin structure regulation, followed by the description of three neurodevelopmental syndromes where altered chromatin structure is believed to be a major causative factor: Cornelia de Lange, Rett and ATR-X syndromes. We highlight common features at the phenotypic and molecular level and discuss the implications for the design of therapies.

2. Basic concepts in chromatin organization and structure

The packaging of DNA into chromatin occurs at different levels. In the primary structure of chromatin, a stretch of DNA is tightly wrapped around four pairs of positively charged structural proteins called histones. Together, the DNA and histones form the basic unit of chromatin known as the nucleosome [6]. In the secondary structure, the nucleosomal array is tightly coiled into a 30 nm chromatin fiber, although whether this arrangement exists *in vivo* has been questioned [7, 8]. The tertiary structure of chromatin consists of higher order chromatin fibre configurations. The density of chromatin packaging, and its dynamic remodelling, affect the accessibility of the DNA to factors involved in DNA replication, transcription and repair [9]. The molecular determinants that influence the level of compaction of chromatin fibres include DNA methylation, nucleosome composition, histone post-translational modifications (PTMs), ATP-dependent chromatin remodelers, and architectural chromatin-associated proteins.

Methylation of DNA in mammals occurs at cytosine residues, in the context of CpG dinucleotides [10]. DNMT3A and DNMT3B are methyltransferases responsible for *de novo* methylation [11]. Accordingly, they are responsible for the wave of *de novo* DNA methylation that occurs in the early embryo [12]. Another enzyme, DNMT1, maintains DNA methylation patterns by acting on newly synthesized DNA to match the parental strand after DNA replication [13-15]. High levels of DNA methylation seem to be correlated with gene inactivity [16]. DNA methylation is involved in gene and transposon silencing [17, 18] and also constitutes the molecular mark that often distinguishes the two alleles of imprinted genes [19-21]. However, DNA methylome analyses in several species have revealed that methylated cytosine residues are also highly enriched in the exons of transcribed genes [22-25]. While the role of exonic DNA methylation is not yet resolved, evidence suggests that it could aid the spliceosome in the process of defining exons [26, 27]. Several derivatives of 5-methylcytosine have now been identified including 5-hydroxymethylcytosine (5-hmc), 5-formylcytosine (5fC) and 5-carboxylcytosine (5caC). They are thought to be generated during the 5-methylcytosine demethylation pathway catalyzed by the Ten-eleven translocation (Tet) enzymes [28-30]. Interestingly, 5-hmc is most abundant in the brain, especially in the hypothalamus and cerebral cortex [31], and its genomic distribution in human and mouse brain showed that it is greatest at synaptic genes and intron-exon boundaries, suggesting an important function in gene splicing and synaptic activity in the central nervous system [32].

The canonical nucleosome consists of pairs of the four core histones H2A, H2B, H3 and H4. Histone H1 binds the DNA between nucleosomes, which is known as linker DNA, and stabilizes higher order chromatin folding [33-36]. During developmental processes such as gene imprinting and X chromosome inactivation, the canonical histones in the nucleosome can be replaced by atypical forms to designate chromosomal regions for specific functions (reviewed in [37, 38]). For instance, the largest of the histones, MacroH2A, acts as a strong transcriptional repressor [39]. It is found in heterochromatin and is associated with the inactive X chromosome in females and the inactive allele of imprinted genes[40-44]. H2A.Z is typically found at transcriptional start sites of active and inactive genes, and is thought to be involved in regulating nucleosome positioning [45-50]. H3.3 incorporation into nucleosomes can remove histone H1 from linker DNA and is thought to facilitate recognition of target sequences by the CCCTC-binding (CTCF) zinc finger protein [51]. Moreover, nucleosomes containing both H2A.Z and H3.3 are particularly labile and often correspond to binding sites for CTCF [52, 53].

Post-translational modifications to core and variant histones are crucial for the dynamic changes in chromatin organization [54, 55]. The amino terminal tails of histones that protrude from the nucleosome core can be marked by methylation, acetylation, ubiquitination and phosphorylation, among a growing list of chemical modifications. These marks are introduced or removed by "writer" and "eraser" proteins, respectively, and are recognized by specific "readers" that alter chromatin properties [54]. For example, histone acetylation is catalyzed by writer enzymes called histone acetyltransferases (HATs) that help open chromatin and correlate with transcriptional activation [56-59]. Conversely, transcription repressor complexes often include histone deacetylases (HDACs) that repress transcription by removing acetyl groups from the histones, promoting a closed chromatin state[60]. Bromodomain-containing proteins are reader proteins that specifically bind to acetylated histones [61]. Methylation marks on histones can be repressive, such as H3K9me3 and H3K27me3 [62-64] or they can denote particular regulatory elements. For example H3K4me3 often flags active gene promoters, which is the start-site of gene transcription [65, 66].

The positioning and density of nucleosomes along the DNA influences many cellular processes, including gene transcription [67]. A shift in nucleosome location can expose regulatory sequences of DNA that contain recognition sites for transcription factors or other regulatory proteins. Nucleosome positioning *in vivo* is dictated by the DNA sequence, the structure of neighbouring chromatin, transcription factors, transcriptional elongation machinery and ATP-dependent chromatin remodeling proteins [67-69]. The displacement of nucleosomes is controlled by protein complexes containing histone chaperones and ATP-dependent nucleosome remodellers. These complexes bind specific histones via the chaperone, and harvest energy from ATP using the ATPase to introduce or displace histones (reviewed in [70]). One example is the Swi/Snf-like ATPase called ATRX that forms a complex with the Daxx histone chaperone to incorporate the histone variant H3.3 into telomeric chromatin [71, 72]. ATRX is known to be involved in human cognition, as mutations in the gene cause intellectual disabilities [73, 74].

Architectural proteins are involved in organizing DNA in the three dimensional space of the cell nucleus [75]. Regulatory sequences like enhancers bind transcription activating protein complexes that interact with distant transcription machinery at the gene promoter through chromatin looping [76, 77]. These chromatin conformations are specific to cell-type and developmental context, as they depend on which transcription and chromatin-associated factors are available in the cell. The activation effect of an enhancer can be blocked by what is known as an insulator sequence. In eukaryotes, CTCF is the only protein known to bind insulator sequences to elicit this blocking effect [78]. CTCF-bound insulators function via the formation of a physically different looping structure, in which the regulatory element can no longer encounter the gene promoter [79, 80]. This type of long-range chromatin fibre interaction also involves the cohesin ring complex, which is believed to encircle DNA strands, as well as ATP-dependent chromatin remodelling proteins [81-83].

Cohesin is a ring complex composed of four proteins: SMC1, SMC3, RAD21 and SA1/SA2 and is genetically conserved from fungi to humans [84]. The ring structure of the complex is formed by interactions between SMC1, SMC3 and RAD21 [85]. The fourth component (either SA1 or SA2) attaches to the ring through interaction with RAD21 and targets cohesin to specific genomic sites [86]. Cohesin and CTCF binding sites largely overlap across the genome, especially near active genes [83, 87]. The current model proposes that CTCF is targeted to its consensus sequence, and cohesin is recruited to the same sites via the SA1/2 subunit [88]. There are now multiple studies demonstrating that cohesin cooperates with CTCF in the formation and stabilization of chromatin looping structures to alter gene expression, including studies at the *H19/Igf2* locus [89], the *IFNG* locus [90], the Beta-globin locus [91], and the MHC class II locus [92]. Depletion of either CTCF or cohesin at these sites results in altered loop formation. In addition, CTCF and cohesin have been demonstrated to mediate interactions of genes and elements on different chromatin fibres [93]. However, cohesin may have a CTCF-independent role in tissue-specific enhancer interactions (reviewed in [94]). A study of murine embryonic stem cells revealed that cohesin localized to a subset of active promoters with a transcriptional coactivator called Mediator, in the absence of CTCF. These genes were expressed in embryonic stem cells through enhancer-promoter interactions that were formed through cohesin-mediator complexes [95]. Cohesin is also required for homologous-recombinational repair of DNA damage following DNA replication [96]. Subunits of cohesin become SUMOylated upon exposure to DNA damaging agents or presence of DNA double-strand breaks by the SUMO E3 ligase Nse2, a subunit of the related Smc5-Smc6 complex [97]. Cohesin was also shown to antagonize binding of the histone variant γH2AX at double-stranded breaks, which may allow for the chromatin remodelling necessary for DNA repair [98].

Normal development and maturation of the human brain relies heavily on the dynamic nature of chromatin and therefore on many of the factors mentioned above. In the next few sections, we discuss particular examples of human disorders with overlapping phenotypes, where mutation of chromatin structure regulators leads to birth defects and intellectual disability (Figure 1).

3. Cornelia de Lange syndrome

Cornelia de Lange Syndrome (CdLS) is a multi-organ developmental disorder characterized by intellectual disability, distinct facial features, growth impairment, short stature and upper limb defects (reviewed in [99]). CdLS causes birth defects in both males and females, and occurs in 1/10,000 to 1/100,000 live births [100, 101]. Clinical manifestations of CdLS range substantially (reviewed in [102, 103]. Facial features and intellectual disability tend to occur in all patients, but limb malformations of the upper extremities present in approximately one third of CdLS patients and range from olidactyly to absent forearm [104]. About one quarter of patients are affected by a congenital heart defect [105-107] or cleft palate [104]. Gastrointestinal abnormalities, diaphragmatic hernia and ambiguous genitalia have also been reported [105, 108, 109].

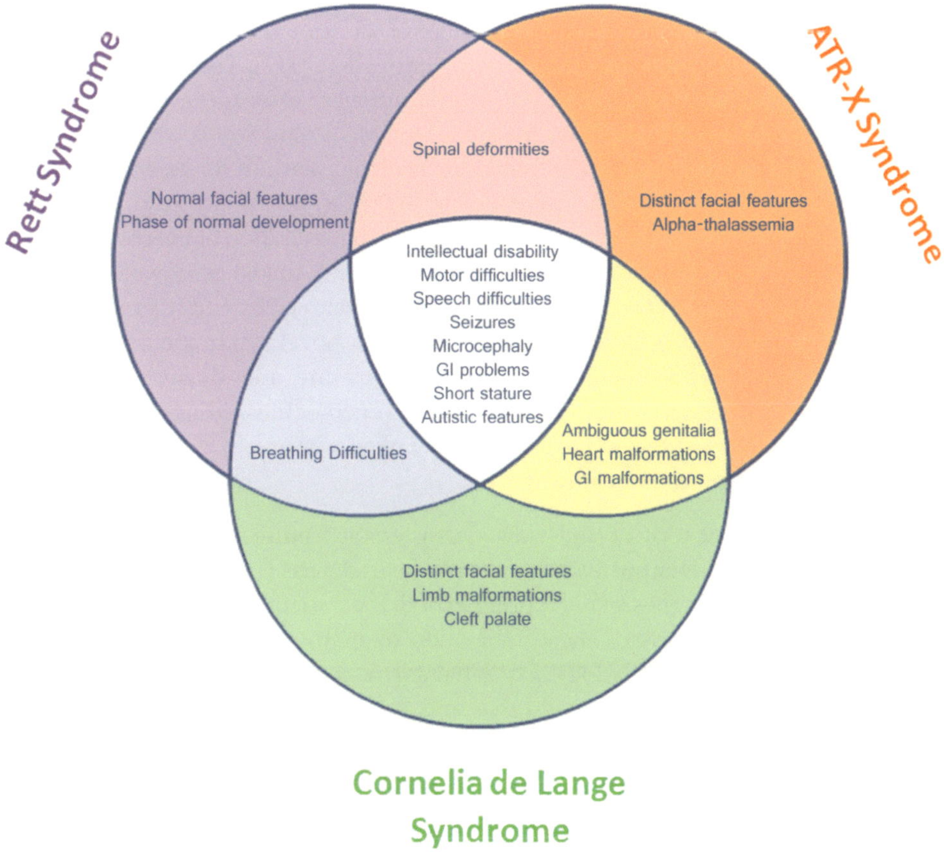

Figure 1. Common clinical features of CdLS, RTT and ATR-X syndromes. Each syndrome is represented by a circle. Features common to two or all three syndromes are listed in the areas of overlap. Multiple clinical features, including Intellectual Disabilities, seizures and microcephaly are shared by all three syndromes.

Central nervous system abnormalities in CdLS include cognitive delay, seizures, self-injurious behaviour, obsessive-compulsive behaviours, attention deficit disorder with or without hyperactivity, and depression. The incidence of structural brain anomalies is unknown, but cerebellar abnormalities have been reported in rare cases [108]. Mild to moderate cases of CdLS are commonly reported to have features of autism [99].

CdLS is caused by mutations in the components of the cohesin complex or its regulatory proteins (reviewed in [84]). Cohesin is responsible for keeping sister chromatids linked during mitosis and meiosis, a process termed sister chromatid cohesion, until they are pulled apart into separating daughter cells (reviewed in [110]). However, abnormal cell division does not satisfyingly explain the molecular cause of CdLS, as only a small fraction of cells from less than half of CdLS patient-derived cell lines show defects in sister chromatid cohesion [111]. Rather, accumulating evidence suggests that a deregulation of gene expression is likely to be the biggest contributor to the symptoms in CdLS [112, 113].

Haploinsufficiency for *NIPPED-B-LIKE* (*NIPBL*) is the most frequent cause of CdLS, with *NIPBL* gene mutations occurring in more than half of all cases [114-116]. NIPBL is a highly conserved protein that facilitates cohesin loading onto DNA [117]. The causative mutations tend to occur *de novo,* and a single mutant allele is sufficient to result in the most common autosomal dominant form of CdLS [118]. Mutations in *SMC1A* and *SMC3,* which belong to the family of structural maintenance of chromosomes proteins, account for an additional 5-10% of CdLS cases [119]. Recently, histone deacetylase 8 (HDAC8) was identified as the vertebrate protein responsible for the deacetylation of SMC3 and the dissolution of the cohesin complex at anaphase [120]. Loss-of-function mutations in *HDAC8*, an X-linked gene, were identified in six of 154 individuals affected by CdLS, including two females [120]. *HDAC8* mutations were also identified in 7 males of a Dutch family affected by a novel syndrome characterized by intellectual disability, hypogonadism, obesity, short stature and distinct facial features reminiscent of Wilson-Turner Syndrome (WTS) [121]. Together the findings from this and the Deardorf et al study suggest that WTS may be an X-linked variant of CdLS, or that CdLS and WTS share a causative molecular pathway.

Mice carrying one mutant copy of *Nipbl* have characteristic features of CdLS including facial anomalies, small size, behavioural disturbances and heart defects [113]. Modest changes in the expression of hundreds of genes were reported in both the mutant mice and in CdLS cell lines [113]. This suggests that perhaps the combination of many small changes in expression culminates into the observed pathology. Experimental manipulation of NIPBL target genes in a zebrafish model indeed revealed additive and synergistic interactions on phenotypic outcomes [112]. Similarly, a lymphoblastic cell line generated from one of the CdLS patients with an *HDAC8* mutation showed that the gene expression profile was strongly correlated with that seen in *NIPBL* mutant cell lines, and not cell lines from control individuals [120]. Gene expression profiling of *Nipbl* mutant embryonic brain tissue revealed a marked down-regulation of the *Protocadherin-beta* (*Pcdh-β*) genes [113]. In mice, the *Pcdh-α*, *-β*, and *-γ* genes are arranged in tandem arrays on chromosome 18 [122]. The clustered *Pcdh* genes comprise >50 putative synaptic recognition molecules that are related to classical cadherins and highly expressed in the nervous system. They are located at both pre- and post-synaptic terminals, making them ideal participants in synapse formation. Only a small subset of protocadherins are expressed in each neuron from the time they are born and a combinatorial effect of protocadherin expression is generated by alternative splicing and promoter usage and is postulated to instruct future synaptic connections and shape the brain's neuronal circuitry [123, 124]. A similar effect *on Pcdh-β* was reported in the SA1-null embryonic brain, and also in CTCF-null pyramidal neurons [113, 125]. SA1 is largely responsible for cohesin accumulation at promoters and at sites bound by CTCF, emphasizing the linkage between these proteins and their importance for normal protocadherin gene expression in the brain [126].

NIPBL might regulate gene expression by controlling loading and unloading of cohesin onto chromatin, thus counteracting its insulating functions [127]. However, it can also recruit

HDAC1 and HDAC3, suggesting that it may promote chromatin remodelling in this way, leading to gene silencing [128]. Genome-wide analysis of DNA methylation in cell lines derived from CdLS patients show specific methylation patterns that differ from controls, specifically on the X chromosome [129]. It is not clear why DNA methylation is affected in CdLS and whether this impacts gene expression changes seen in the disorder.

4. Rett syndrome (RTT)

Rett syndrome (RTT) is a neurodevelopmental disorder characterized by intellectual disability, autistic features, increased risk of epilepsy, and a loss of previously achieved motor and language milestones. RTT affects about one girl in 10,000-15,000, making it the second leading cause of intellectual disability in females, after Down syndrome [130]. In 1999, Amir et al found that RTT was caused by mutations of the Methyl-CpG-binding protein 2 gene (*MeCP2*) [131]. RTT is mainly sporadic and the majority of mutations appear to be of paternal origin. Mutations alter protein sequence or result in truncated versions of MeCP2 with residual function [131]. Males carry only a single copy of the *MeCP2* gene due to its location on the X chromosome. Since at least one functional copy of the gene is required, males with mutations in *MeCP2* are rarely affected with RTT but rather exhibit severe encephalopathy [132-134]. In a typical course of the disease, RTT patients experience normal development up to 6-18 months of age, followed by a period of arrested developmental progress and eventual regression with poor social contact and finger skills (reviewed in 135). In early childhood, the majority of patients have gastrointestinal problems including difficulty swallowing, which likely contributes to malnutrition and pervasive growth problems [136]. In addition, about half of patients have head circumferences below the 3rd percentile (microcephaly), curvature of the spine (scoliosis), and are unable to walk [137].

Much debate still surrounds the question of MeCP2 function at the molecular level, perhaps due to the confounding effects of various post-translational modifications of the protein. MeCP2 is an intrinsically disordered protein that binds methylated DNA via the methyl-binding domain (MBD). One of the roles ascribed to MeCP2 is that of a transcription repressor that binds methylated gene promoters and recruits repressive factors including HDAC1, HDAC2 and Sin3A [138]. For example, recruitment of this complex by MeCP2 regulates the expression of brain-derived neurotrophic factor (BDNF), a protein with important roles in neuronal survival and synaptic plasticity (reviewed in [139]). Neuronal activity leads to demethylation of the *Bdnf* gene, dissociation of the MeCP2-HDAC complex, and increased gene transcription [140, 141].

Despite this somewhat satisfying and simple model of MeCP2 function as a transcriptional repressor, identification of over-expressed target genes in MeCP2-deficient tissue has been difficult. Even more unsettling was the discovery that half of the genes with MeCP2 bound within the promoter in wild type brain actually showed decreased expression in MeCP2-null tissues [142]. The explanation for these discrepancies may come from emerging data indicating that MeCP2 may regulate the organization and compaction of chromatin at a more global level.

A study by Skene et al. proposed that MeCP2 does not act only in a locus-specific manner, but displays a histone-like distribution across the genome in neurons [143]. Moreover, they found large-scale chromatin changes in neurons of MeCP2-null mice, including elevated histone H3 acetylation and doubling of histone H1 in chromatin. Supporting data comes from *in vitro* studies of MeCP2, showing that it can induce compaction-related changes in nucleosome architecture that resemble the classical zigzag motif induced by histone H1 and considered important for 30-nm-fiber formation. The doubling of histone H1 in MeCP2-null neurons may be explained by the finding of Ghosh et al, which suggests that MeCP2 competes with H1 for common binding sites [144]. Consistent with a broader role in chromatin structure organization, MeCP2 is homologous to the attachment region binding protein (*Arbp*) gene in chicken, which has roles in chromatin looping [145]. ARBP has high affinity for specific DNA sequences known as MAR/SARs which it organizes onto a nuclear matrix scaffold [146]. This suggests that ARBP, and by extension perhaps also MeCP2, is involved in chromatin loop organization. MeCP2 loss-of-function was indeed shown to rearrange chromatin fibre interactions at the *Dlx5* locus in mouse brain cells using the chromatin conformation capture technique [147]. These results may be highly relevant to RTT pathology given that the DLX5 protein is an important regulator of GABAergic interneuron development [148]. GABAergic signalling plays a vital role in modulating the activity of the cerebral cortex, and alterations in interneuron position and/or migration have been linked to mental retardation, autism, schizophrenia, epilepsy and Down syndrome [149]. Two GABAergic interneuron-specific MeCP2 knockout mouse lines were generated that exhibited reduced GABA levels in their cortices and displayed repetitive behaviours reminiscent of RTT, including hindlimb clasping, forelimb stereotypies and over-grooming leading to fur loss. In addition, these GABA-specific MeCP2 knockout mice showed progressive motor dysfunction [150].

The methyl binding domain (MBD) of MeCP2 targets the protein to methylated DNA and allows for clustering of pericentric heterochromatin *in vivo* [151]. Analysis of 21 RTT patient mutations showed that two thirds of these decreased the ability of MeCP2 to cluster heterochromatin in mouse cells [152]. This led to the question whether heterochromatin aggregation is impaired in these mutants because of the inability to bind methylated DNA or because of a different function of MeCP2. MeCP2 has multiple chromatin-interacting domains as well as a methylation-independent DNA binding domain *in vitro* [153]. There was some evidence suggesting that the ability of MeCP2 to control chromatin condensation did not require methylated DNA [147, 154, 155], but it was a study by Casas-Delucchi et al that demonstrated that the role of MeCP2 in heterochromatin condensation was independent of DNA binding [156]. They designed an assay in which different mutant MeCP2 proteins from RTT patients were artificially targeted to heterochromatic regions in living cells by fusion to a heterochromatin-binding protein. This allowed for the effects of MeCP2 mutants on chromatin dynamics and organization to be observed *in vivo*. Some RTT mutations led to exclusive decreases in methylated DNA binding, without influencing the ability of MeCP2 to cluster heterochromatin, while other mutations affected both functions. In those mutants that were able to cluster heterochromatin, fusion of large heterochromatic structures (over several micrometers in size) were visualized *in vivo*, providing evidence for the ability of MeCP2 to mediate large-scale chromatin rearrangements [156].

Several mutant mouse models have been generated to study the effects of MeCP2 deficiency (reviewed in [157]). Each RTT model strain has a slightly different time of adverse phenotype onset, but the males usually begin displaying abnormal behaviours between 4-6 weeks after birth. The defects observed in these strains of mice generally recapitulate the symptoms of RTT female patients: laboured breathing, reduced exploratory activity, seizures, cognitive deficits and decreased synaptic plasticity [157-160]. RTT mice clasp their hind paws when suspended by the tail, which is a common sign of neurological deficits [161]. They also have decreased brain weight, smaller cortical neurons with increased neuronal cell density, and reduced dendritic arborisation compared with controls [162, 163]. Importantly, transgenic mice that overexpress MeCP2 also exhibit behaviours of anxiety and impairments in learning and memory, demonstrating that neurons are highly susceptible to either decreased or increased levels of MeCP2.

Several studies provide clues as to the cause of intellectual disability seen in RTT patients. Some of the findings show deficits in long-term potentiation and long-term depression and reduction in spontaneous neurotransmission in cortical and hippocampal neurons of Mecp2-null mice [164-167]. Furthermore, post-mortem tissue displays immature neuronal dendrite morphology predicted to result from altered synaptic activity [162, 168, 169].

5. Alpha-Thalassemia, mental Retardation X-linked syndrome (ATR-X)

ATR-X syndrome is a rare genetic disorder characterized by moderate to severe intellectual and motor disability, mild alpha-thalassemia in a subset of cases, as well as specific developmental abnormalities including facial, skeletal and urogenital defects [170, 171]. ATR-X syndrome affects very few individuals as it frequently results from familial mutations in the *ATRX* gene on the X chromosome that are passed on to sons by carrier females. In 2009, there were over 200 known male patients [172]. Females are rarely affected due to skewed X chromosome inactivation, in which the X chromosome carrying the mutant *ATRX* gene is preferentially selected for condensation [173, 174].

Manifestation of ATR-X syndrome can be quite variable [175, 176]. Typically, affected males have severe global delay from birth, developing very little language and motor abilities. Approximately one third of patients have seizures, and microcephaly is not uncommon (reviewed in [177]). These patients also often have gastrointestinal abnormalities, including difficulty swallowing and gastro-esophageal reflux, which has been known to cause death by asphyxiation in multiple ATR-X cases [178]. Some patients have anatomical abnormalities that can cause stomach torsion or cause severe constipation [178]. Genital abnormalities exist in about 80% of cases, and can present as undescended testes, hypospadias, ambiguous genitalia or normal female genitalia [177].

ATR-X syndrome is caused by mutations in the X-linked ATP-dependent chromatin remodelling protein called ATRX [73]. Mutations in the *ATRX* gene have also been identified in previously characterized mental retardation disorders: Juberg-Marsidi syndrome [179], Carpenter-Waziri syndrome [180], Smith-Fineman-Myers [176] and X-linked mental retardation with spastic paraplegia [181]. These disorders were mistakenly thought to be distinct from ATR-X syndrome, as there were mild differences in patient presentations. Identification of *ATRX* mutations in these cases exemplifies the clinical variation that can occur in ATR-X syndrome.

The *ATRX* mutations identified to date alter protein sequence or code for truncated forms of ATRX, resulting in reduced protein function or protein level [73, 182]. Nearly all of these mutations are found within the two functional domains of ATRX, located at the end termini of the protein [73, 182, 183]. The domain located at the N-terminus is known as the ADD (ATRX-DNMT3-DNMT3L) domain [184]. It consists of DNA-binding zinc fingers, a protein-binding plant homeodomain finger, and a globular region [185]. The ADD domain is a histone H3-binding module that is selective for the combinatorial readout of H3K9 trimethylation and the lack of H3K4 trimethylation [186, 187]. The domain located at the C-terminus displays ATPase and helicase activity, and is homologous to protein regions found in Swi2 / Snf2 family members. This region allows for Swi2/Snf2 proteins to modulate histone-DNA interactions using energy from ATP hydrolysis [188]. ATRX protein-interactions are consistent with a role in chromatin regulation as ATRX has been shown to interact with HP1α [189, 190], EZH2 [191], Mecp2 [155, 192], Daxx [193, 194] and cohesin[155]. Together, the domain analyses and protein interactions of ATRX suggest a role in ATP-dependent alteration of chromatin. ATRX interacts with EZH2 at repetitive sites in centromeres, telomeres and at ribosomal DNA to control heterochromatin formation [195-197]. Heterochromatin formation is further induced by the interaction of ATRX with HP1, a protein that functions in binding and maintaining hetero-chromatic marks like trimethylated lysine 9 of histone 3 (H3K9Me3) and trimethylated lysine 20 of histone 4 (H4K20Me3) [198].

Forebrain-specific deletion of ATRX causes increased p53-dependent neuronal apoptosis, resulting in reduced forebrain size and hypocellularity of the cortex and hippocampus [199, 200]. Many of the mutant mice die in the neonatal period of unknown causes. In contrast, mice expressing a truncated form of ATRX (ATRX(ΔE2) mice) survive and reproduce normally [201]. Behavioural analyses of these mice showed that they have defects in contextual fear memory with dysfunction of calcium/ calmodulin-dependent protein kinase II (CaMKII) and GluR1 [201]. Further studies showed abnormally increased CamKII activity in the prefrontal cortex of the ATRX(ΔE2) mice [202]. In addition, their prefrontal cortex contained neurons with longer and thinner dendritic spines than those found in controls, which is consistent with other mouse models of intellectual disabilities [202].

Previous work has also shown that genes are deregulated in cells of ATR-X patients and ATRX mutant mice [203, 204]. Two possible mechanisms by which ATRX can act as a transcriptional regulator have been demonstrated [193, 205]. The presence of Daxx relieves the repressive effect of ATRX, but not through alteration of its ATPase activity [193]. This is now understood to occur through the role of Daxx as a chaperone for histone variant H3.3, a marker of active chromatin. Daxx assists in H3.3-H4 tetramer deposition at nucleosomes at PML nuclear bodies, ribosomal DNA, pericentric DNA and telomeres [71, 206]. One theory is that ATRX directs Daxx to deposit H3.3 at specific chromatin regions that have been made accessible by ATRX through ATP-dependent remodelling. ATRX also acts as an inhibitor of macroH2A deposition into chromatin [205]. In ATRX-null cells, macroH2A accumulates at the HBA gene cluster and leads to reduced α-globin expression [205]. This is thought to contribute to the symptom of α-thalassemia seen in ATR-X syndrome patients.

The mechanism by which ATRX may be able to direct Daxx to specific sites is unknown. One possibility is that ATRX localizes to specific loci through the ADD domain [186, 187, 207]. The ADD domain of ATRX was shown to contain two binding pockets for histone 3 modifications: one for unmodified lysine 4 and the other for trimethylated lysine 9 [207]. The combination of these two histone 3 marks is associated with heterochromatin / silent gene promoters and methylated DNA (reviewed in [208]).This combinatorial binding is required for ATRX localization *in vivo* [207]. Further, mutations in *ATRX* that disrupt the interaction of the ADD domain with H3K9me3 cause a loss of ATRX targeting to heterochromatin [187].

The localization of ATRX to H3K9me3 is strengthened by interaction with HP1α, which also binds H3K9me3 [207]. In addition, ATRX has been shown to be recruited by Mecp2 [192], which binds the methylated DNA associated with these histone modifications. In fact, loss of MeCP2 in mice results in a loss of ATRX localization at heterochromatic sites in neurons [192]. In addition, a subset of RTT patient *MeCP2* mutations interfere with ATRX-MeCP2 interaction [192], which suggests that RTT can be caused in some cases by the inability of MeCP2 to recruit ATRX to specific chromatin sites.

Recently, a family was identified with two men affected by concomitant duplication of both *Mecp2* and *ATRX* [209]. These men did not exhibit signs of ATRX duplication syndrome (short stature, and hypoplastic genitalia), but instead presented with severe mental retardation, muscular hypotonia, and other characteristic features of MeCP2 duplication syndrome. This finding supports the idea that MeCP2 acts upstream of ATRX. However, there was an added feature (cerebellar atrophy) in these patients that was inconsistent with Mecp2 duplication syndrome, which suggests that ATRX may have some additive effect, and not always function in a pathway with Mecp2.

We previously reported that ATRX, MeCP2 and cohesin might cooperate in transcriptional regulation in the brain [155]. In this study, we utilized mice that lack the ATRX protein specifically in the forebrain, by Cre-loxP recombination [199]. These mice have reduced cortical and hippocampal size, reduced number of GABAergic interneurons and exhibit gene expression changes [199, 200, 203]. In control mice, we could show that ATRX and MeCP2 localize to the maternal allele of the *H19* imprinted gene at the upstream imprinting control region (*H19* ICR). In the absence of ATRX, *H19* gene repression in the postnatal period was lessened and correlated with reduced occupancy of cohesin and CTCF at the *H19* ICR. These findings suggest that ATRX is required for optimal gene repression through the recruitment of CTCF and cohesin or by promoting their stable binding to chromatin. A link between ATRX and chromatin cohesion was not only found in the context of gene regulation, but also during mitosis and meiosis. Depletion of ATRX protein in human somatic cells resulted in several mitotic defects, such as mis-congression, reduced cohesion and condensation, mis-segregation of chromosomes and the formation of micronuclei [210]. Abnormal chromosome congression and segregation may in part explain the reduced brain size of forebrain-specific ATRX knockout mice [210].

Genome wide assessment of ATRX protein binding was performed in mouse embryonic stem cells and human erythroblast cells [195]. ATRX binding was often seen at high GC-rich regions of the genome, including the telomeres. These DNA sequences have a high probability of forming unusual DNA structures called G-quadruplexes, or G4-DNA, and recombinant ATRX protein was able to bind these structures *in vitro*. G-quadruplex structures are believed to influence many cellular processes such as transcription elongation and DNA replication and could prove to be an important feature in understanding CNS defects caused by the loss of ATRX protein activity.

6. Therapeutic implications

The shared phenotypic features of CdLS, RTT and ATR-X syndrome (Figure 1) in combination with the molecular findings that place cohesin, MeCP2 and ATRX together in the same physical and functional context (Figure 2) suggest that these three syndromes are in part due to aberration of the same molecular pathways. In particular, the shared feature of intellectual disability and the joint role of MeCP2, ATRX and cohesin in chromatin organization demonstrate that the regulation of chromatin structure is essential for the development of the brain and its complex functions. The study of chromatin structure regulation in the brain, and the identification of defects in gene expression that are caused as a result of abnormal chromatin organization, have been valuable not only to our understanding of human syndromes, but also to the development of therapeutics. This has been especially true of MeCP2 and RTT, the most studied of the three syndromes.

One interesting feature of RTT is that MeCP2-null neurons in the brain do not undergo programmed cell death, or apoptosis[211]. In fact, mounting evidence suggest that RTT is not a neurodegenerative disease, but rather a disorder of neuronal activity (reviewed in [212]). The changes in synaptic maturation and neuronal activity are in part a result of impaired chromatin regulation. Chromatin modifications are dynamic and reversible, which led to the hypothesis that RTT defects may be reversible as well. In 2007, Guy et al. demonstrated that activation of MeCP2 expression in adult MeCP2-deficient mice, even at an advanced stage of illness, reversed neurological symptoms [213, 214]. Replication studies have also shown reversal of RTT morphological features, including neuronal size and dendritic complexity, as well as improvement in functional RTT symptoms such as respiratory function, grip strength and rotarod performance, with the reactivation of MeCP2 in mice [215]. These results have since revolutionized the way in which intellectual disability syndromes are understood [212, 216].

However, there are many obstacles for which gene therapy cannot currently be considered in RTT patients (reviewed in [217]). Gene dosage is one important consideration; since MeCP2 is located on the X chromosome, the number of neurons affected in each patient is dependent on X-inactivation. Providing excess MeCP2 to neurons that already express the non-mutant allele has negative consequences on brain function. This has been observed in mice [214], as well as in humans where severe intellectual disability caused by MeCP2 duplication has been documented (MeCP2 Duplication Syndrome [134, 218, 219]). Therefore a specific dose of the

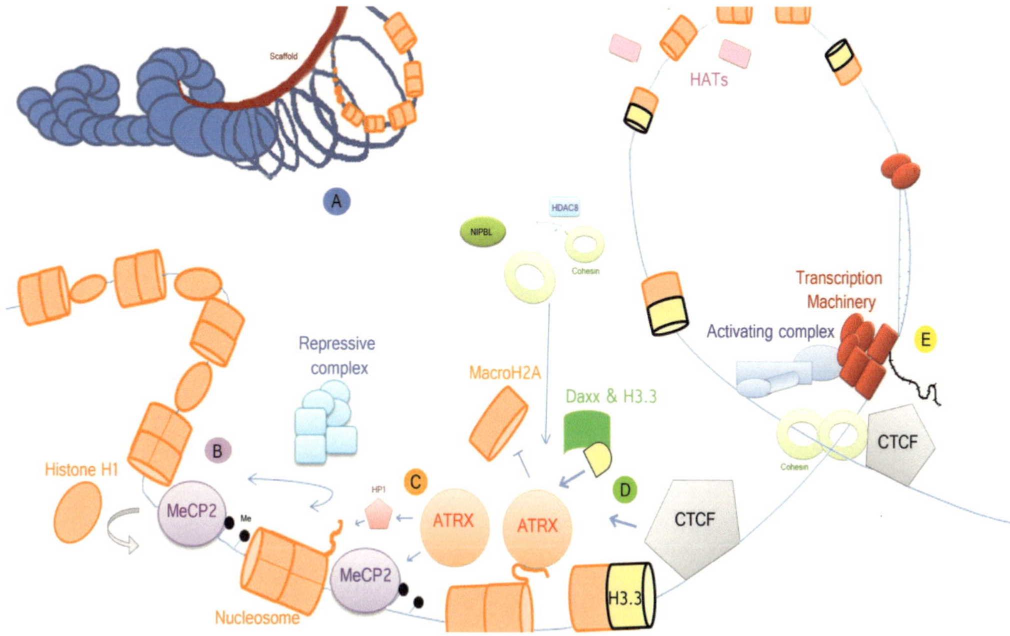

Figure 2. Regulation of chromatin organization by cohesin, ATRX and MeCP2. A: DNA is wrapped around histones in a complex known as the nucleosome. The nucleosome-covered DNA is coiled to form a 30 nm fibre which then further coils and loops to form higher order structures. These chromatin structures are attached to scaffolds in the nucleus. B: MeCP2 competes for linker DNA with histone H1 at methylated cytosine residues. MeCP2 recruits repressive complexes which contain HDACs that deacetylate histone tails. Unacetylated histone 3 Lysine 9 becomes trimethylated and attracts HP1alpha. C: ATRX is recruited by MeCP2. MeCP2 and HP1alpha both directly interact with ATRX. D: ATRX has binding sites for H3K9me3 and unmodified H3K4. ATRX inhibits macroH2A incorporation into nucleosomes and recruits Daxx, which is a chaperone for Histone H3.3. Histone H3.3 is incorporated into nucleosomes and marks active chromatin. ATRX recruits CTCF and cohesin. Cohesin is loaded onto DNA by NIPBL. E: Cohesin and CTCF interact to stabilize looping structures. These loops allow transcription machinery to interact with distant activating complexes bound to enhancers.

gene is required. This issue is not unique to MeCP2; over-expression of ATRX in mice led to disorganization of the cells in the brain at the ventricular zone, seizures and death soon after birth [220]. Case studies in humans report that ATRX duplication is associated with severe intellectual disability, genital anomalies and short stature [221, 222].

Due to these and other issues, therapeutic approaches in RTT have had to focus on pathways downstream of MeCP2. The understanding of how MeCP2 perturbs gene expression through its effects on chromatin has been indispensable to these advances. For example, a link between MeCP2 and the regulation of Brain-derived neurotrophic factor (BDNF) expression led Tsai et al. to test the administration of BDNF on the phenotypic outcomes of MeCP2 mutant mice [223, 224]. BDNF is a secreted factor of the neurotrophin family that promotes survival of neurons but also growth and differentiation of new neurons and synapses. BDNF injection led to a slower progression of disease in the RTT mouse model. Potentially, intravenous injection of BDNF in RTT patients could increase BDNF levels in the brain and slow the progression of symptoms. In particular, breathing dysfunction leads to increased mortality and morbidity in

RTT. Reduced levels BDNF in the brain of mice is associated with increased tachypneas and apneas [225-228]. Pharmacological activation of the BDNF receptor TrkB in RTT mice restored wild-type breathing, which demonstrates another potential avenue for therapy in RTT [225]. However, the treatment with the most promise in current literature is the administration of Insulin-like Growth Factor 1 (IGF-1). IGF-1 is an important regulator of synaptic plasticity and maturation that is widely expressed in the brain (reviewed in [229]). Multiple studies have supported the hypothesis that dendritic spines are altered in RTT, implicating synaptic maturation as a major deficit. Treatment of RTT model mice with IGF-1 N-terminal tripeptide, known as GPE, partially restores dendrite spine number, and improves the cortical plasticity levels to that of wild type mice [230]. In addition, it improves gait and breathing patterns of the MeCP2 mutant mice.

7. Conclusions

Animal models have provided an extensive knowledge about the three syndromes discussed above. However these animal studies cannot recapitulate all of the complexities of human brain disorders. Neurodevelopmental disorders have been difficult to study in humans because of the limited supply of post-mortem brain samples and studying peripheral cells from patients such as lymphocytes is problematic because they do not accurately portray defects of the target tissue [231]. Human induced pluripotent stem cells (iPSCs) are a novel technology that may provide a potential solution to this issue. iPSCs are a type of stem cell that are produced by genetic reprogramming of a differentiated somatic cell [232, 233]. These iPSCs can be derived from healthy individuals or from those afflicted by a genetic condition, and then differentiated into the cell type desired for research. Studies of neuronal cells derived from iPSCs of RTT patients have provided valuable complimentary information to the findings from *in vivo* animal studies. Specifically, modeling RTT with iPSCs has allowed for medications like IGF1 to be tested for efficacy in human RTT patient neurons [234]. Administration of IGF1 was shown to rescue synaptic defects in this model and is currently in clinical trials for treatment of RTT for which primary outcome measures will be available in 2013 (http://clinicaltrials.gov/ct2/show/record/NCT01253317?term=rett+syndrome). It has not been shown whether IGF1 is affected by ATRX or NIPBL knockdown, or whether similar therapies would be beneficial in ATR-X or Cornelia de Lange syndromes. However, since there is evidence that ATRX, cohesin and MeCP2 function together in regulating gene expression and brain development, it is possible that downstream targets, like IGF1, are similarly affected in all three syndromes.

Although more work remains, the study of chromatin modifiers in brain development have provided insight into inherited forms of intellectual disabilities, as well as target pathways for future clinical interventions. Continued investigation of chromatin regulation in neurological and psychiatric disease will help to identify more commonalities between disorders and further our knowledge of potential treatment avenues.

Acknowledgements

We wish to acknowledge funding for this work from the Canadian Institutes for Health Research (CIHR; MOP93697).

A.E. is the recipient of a CIHR Vanier Scholarship.

Author details

Adrienne Elbert[1,3] and Nathalie G. Bérubé[1,2,3]

*Address all correspondence to: nberube@uwo.ca

1 Children's Health Research Institute and Department of Paediatrics, Western University. Victoria Research Laboratories, London, Canada

2 Department Biochemistry, Western University. Victoria Research Laboratories, London, Canada

3 Schulich School of Medicine and Dentistry, Western University. Victoria Research Laboratories, London, Canada

References

[1] Lyons, M. R, & West, A. E. (2011). Mechanisms of specificity in neuronal activity-regulated gene transcription. *Prog Neurobiol* , 94, 259-295.

[2] Hu, X. L, Wang, Y, & Shen, Q. (2012). Epigenetic control on cell fate choice in neural stem cells. *Protein Cell* , 3, 278-290.

[3] Leeb, M, & Wutz, A. (2012). Establishment of epigenetic patterns in development. *Chromosoma* , 121, 251-262.

[4] Alabert, C, & Groth, A. (2012). Chromatin replication and epigenome maintenance. *Nat Rev Mol Cell Biol* , 13, 153-167.

[5] Brunner, A.M., Tweedie-Cullen, R.Y., and Mansuy, I.M. 2012. Epigenetic modifications of the neuroproteome. Proteomics 12:2404-2420

[6] Kornberg, R. D. (1974). Chromatin structure: a repeating unit of histones and DNA. *Science* , 184, 868-871.

[7] Joti, Y, Hikima, T, Nishino, Y, Kamda, F, Hihara, S, Takata, H, Ishikawa, T, & Maeshima, K. (2012). Chromosomes without a nm chromatin fiber. *Nucleus* 3., 30.

[8] Bian, Q, & Belmont, A. S. (2012). Revisiting higher-order and large-scale chromatin organization. *Curr Opin Cell Biol* , 24, 359-366.

[9] Jackson, V. (1990). In vivo studies on the dynamics of histone-DNA interaction: evidence for nucleosome dissolution during replication and transcription and a low level of dissolution independent of both. *Biochemistry* , 29, 719-731.

[10] Bird, A. P. (1986). CpG-rich islands and the function of DNA methylation. *Nature* , 321, 209-213.

[11] Lyko, F, Ramsahoye, B. H, Kashevsky, H, Tudor, M, Mastrangelo, M. A, Orr-weaver, T. L, & Jaenisch, R. (1999). Mammalian (cytosine-5) methyltransferases cause genomic DNA methylation and lethality in Drosophila. *Nat Genet* , 23, 363-366.

[12] Clouaire, T, & Stancheva, I. (2008). Methyl-CpG binding proteins: specialized transcriptional repressors or structural components of chromatin? *Cell Mol Life Sci* , 65, 1509-1522.

[13] Li, E, Bestor, T. H, & Jaenisch, R. (1992). Targeted mutation of the DNA methyltransferase gene results in embryonic lethality. *Cell* , 69, 915-926.

[14] Gruenbaum, Y, Cedar, H, & Razin, A. (1982). Substrate and sequence specificity of a eukaryotic DNA methylase. *Nature* , 295, 620-622.

[15] Bestor, T. H, & Ingram, V. M. (1983). Two DNA methyltransferases from murine erythroleukemia cells: purification, sequence specificity, and mode of interaction with DNA. *Proc Natl Acad Sci U S A* , 80, 5559-5563.

[16] Yeivin, A, & Razin, A. (1993). Gene methylation patterns and expression. *EXS* , 64, 523-568.

[17] Wu, H, & Zhang, Y. (2011). Mechanisms and functions of Tet protein-mediated 5-methylcytosine oxidation. *Genes Dev* , 25, 2436-2452.

[18] Wu, H, Alessio, D, Ito, A. C, Wang, S, Cui, Z, Zhao, K, Sun, K, Zhang, Y. E, & Genome-wide, Y. analysis of 5-hydroxymethylcytosine distribution reveals its dual function in transcriptional regulation in mouse embryonic stem cells. *Genes Dev* , 25, 679-684.

[19] Reik, W, Collick, A, Norris, M. L, Barton, S. C, & Surani, M. A. (1987). Genomic imprinting determines methylation of parental alleles in transgenic mice. *Nature* , 328, 248-251.

[20] Swain, J. L, Stewart, T. A, & Leder, P. (1987). Parental legacy determines methylation and expression of an autosomal transgene: a molecular mechanism for parental imprinting. *Cell* , 50, 719-727.

[21] Chaillet, J. R, Vogt, T. F, Beier, D. R, & Leder, P. (1991). Parental-specific methylation of an imprinted transgene is established during gametogenesis and progressively changes during embryogenesis. *Cell* , 66, 77-83.

[22] Choi, J. K. (2010). Contrasting chromatin organization of CpG islands and exons in the human genome. *Genome Biol* 11:R70.

[23] Anastasiadou, C, Malousi, A, Maglaveras, N, & Kouidou, S. (2011). Human epigenome data reveal increased CpG methylation in alternatively spliced sites and putative exonic splicing enhancers. *DNA Cell Biol* , 30, 267-275.

[24] Flores, K. B, Wolschin, F, Allen, A. N, Corneveaux, J. J, Huentelman, M, & Amdam, G. V. (2012). Genome-wide association between DNA methylation and alternative splicing in an invertebrate. *BMC Genomics* 13:480.

[25] Bonasio, R, Li, Q, Lian, J, Mutti, N. S, Jin, L, Zhao, H, Zhang, P, Wen, P, Xiang, H, Ding, Y, et al. (2012). Genome-wide and Caste-Specific DNA Methylomes of the Ants Camponotus floridanus and Harpegnathos saltator. *Curr Biol.*

[26] Malousi, A, & Kouidou, S. (2012). DNA hypermethylation of alternatively spliced and repeat sequences in humans. *Mol Genet Genomics* , 287, 631-642.

[27] Oberdoerffer, S. (2012). A conserved role for intragenic DNA methylation in alternative pre-mRNA splicing. *Transcription* , 3, 106-109.

[28] Pfaffeneder, T, Hackner, B, Truss, M, Munzel, M, Muller, M, Deiml, C. A, Hagemeier, C, & Carell, T. (2011). The discovery of 5-formylcytosine in embryonic stem cell DNA. *Angew Chem Int Ed Engl* , 50, 7008-7012.

[29] He, Y. F, Li, B. Z, Li, Z, Liu, P, Wang, Y, Tang, Q, Ding, J, Jia, Y, Chen, Z, Li, L, et al. (2011). Tet-mediated formation of 5-carboxylcytosine and its excision by TDG in mammalian DNA. *Science* , 333, 1303-1307.

[30] Ito, S, Shen, L, Dai, Q, Wu, S. C, Collins, L. B, Swenberg, J. A, He, C, & Zhang, Y. (2011). Tet proteins can convert 5-methylcytosine to 5-formylcytosine and 5-carboxylcytosine. *Science* , 333, 1300-1303.

[31] Munzel, M, Globisch, D, Bruckl, T, Wagner, M, Welzmiller, V, Michalakis, S, Muller, M, Biel, M, & Carell, T. (2010). Quantification of the sixth DNA base hydroxymethylcytosine in the brain. *Angew Chem Int Ed Engl* , 49, 5375-5377.

[32] Khare, T, Pai, S, Koncevicius, K, Pal, M, Kriukiene, E, Liutkeviciute, Z, Irimia, M, Jia, P, Ptak, C, Xia, M, et al. (2012). hmC in the brain is abundant in synaptic genes and shows differences at the exon-intron boundary. *Nat Struct Mol Biol.*, 5.

[33] Shaw, B. R, Herman, T. M, Kovacic, R. T, Beaudreau, G. S, & Van Holde, K. E. (1976). Analysis of subunit organization in chicken erythrocyte chromatin. *Proc Natl Acad Sci U S A* , 73, 505-509.

[34] Whitlock, J. P. Jr., and Simpson, R.T. (1976). Removal of histone H1 exposes a fifty base pair DNA segment between nucleosomes. *Biochemistry* , 15, 3307-3314.

[35] Kornberg, R. D. (1977). Structure of chromatin. *Annu Rev Biochem* , 46, 931-954.

[36] Worcel, A, & Benyajati, C. (1977). Higher order coiling of DNA in chromatin. *Cell* , 12, 83-100.

[37] Gamble, M. J, & Kraus, W. L. (2010). Multiple facets of the unique histone variant macroH2A: from genomics to cell biology. *Cell Cycle* , 9, 2568-2574.

[38] Millau, J. F, & Gaudreau, L. (2011). CTCF, cohesin, and histone variants: connecting the genome. *Biochem Cell Biol* , 89, 505-513.

[39] Doyen, C. M, An, W, Angelov, D, Bondarenko, V, Mietton, F, Studitsky, V. M, Hamiche, A, Roeder, R. G, Bouvet, P, & Dimitrov, S. (2006). Mechanism of polymerase II transcription repression by the histone variant macroH2A. *Mol Cell Biol* , 26, 1156-1164.

[40] Pehrson, J.R, Fried, V.A, & Macro, . 2A, a core histone containing a large nonhistone region. *Science* 257:1398-1400.

[41] Costanzi, C, & Pehrson, J. R. (1998). Histone macroH2A1 is concentrated in the inactive X chromosome of female mammals. *Nature* , 393, 599-601.

[42] Costanzi, C, Stein, P, Worrad, D. M, Schultz, R. M, & Pehrson, J. R. (2000). Histone macroH2A1 is concentrated in the inactive X chromosome of female preimplantation mouse embryos. *Development* , 127, 2283-2289.

[43] Choo, J.H, Kim, J.D, Kim, J, & Macro, . 2A1 knockdown effects on the Peg3 imprinted domain. *BMC Genomics* 8:479.

[44] Choo, J. H, Kim, J. D, Chung, J. H, Stubbs, L, & Kim, J. (2006). Allele-specific deposition of macroH2A1 in imprinting control regions. *Hum Mol Genet* , 15, 717-724.

[45] Marques, M, Laflamme, L, Gervais, A. L, & Gaudreau, L. (2010). Reconciling the positive and negative roles of histone H2A.Z in gene transcription. *Epigenetics* , 5, 267-272.

[46] Fan, J. Y, Gordon, F, Luger, K, Hansen, J. C, & Tremethick, D. J. (2002). The essential histone variant H2A.Z regulates the equilibrium between different chromatin conformational states. *Nat Struct Biol* , 9, 172-176.

[47] Gevry, N, Hardy, S, Jacques, P. E, Laflamme, L, Svotelis, A, Robert, F, & Gaudreau, L. A.Z is essential for estrogen receptor signaling. *Genes Dev* , 23, 1522-1533.

[48] Guillemette, B, Bataille, A. R, Gevry, N, Adam, M, Blanchette, M, Robert, F, & Gaudreau, L. (2005). Variant histone H2A.Z is globally localized to the promoters of inactive yeast genes and regulates nucleosome positioning. *PLoS Biol* 3:e384.

[49] Kumar, S. V, & Wigge, P. A. (2010). H2A.Z-containing nucleosomes mediate the thermosensory response in Arabidopsis. *Cell* , 140, 136-147.

[50] Thakar, A, Gupta, P, Ishibashi, T, Finn, R, Silva-moreno, B, Uchiyama, S, Fukui, K, Tomschik, M, Ausio, J, & Zlatanova, J. (2009). H2A.Z and H3.3 histone variants affect

nucleosome structure: biochemical and biophysical studies. *Biochemistry* , 48, 10852-10857.

[51] Braunschweig, U, Hogan, G. J, Pagie, L, & Van Steensel, B. binding is inhibited by histone variant H3.3. *EMBO J* , 28, 3635-3645.

[52] Jin, C, Zang, C, Wei, G, Cui, K, Peng, W, Zhao, K, & Felsenfeld, G. (2009). H3.3/ H2A.Z double variant-containing nucleosomes mark'nucleosome-free regions' of active promoters and other regulatory regions. *Nat Genet* , 41, 941-945.

[53] Fu, Y, Sinha, M, Peterson, C. L, & Weng, Z. (2008). The insulator binding protein CTCF positions 20 nucleosomes around its binding sites across the human genome. *PLoS Genet* 4:e1000138.

[54] Strahl, B. D, & Allis, C. D. (2000). The language of covalent histone modifications. *Nature* , 403, 41-45.

[55] Jenuwein, T, & Allis, C. D. (2001). Translating the histone code. *Science* , 293, 1074-1080.

[56] Imhof, A, Yang, X. J, Ogryzko, V. V, Nakatani, Y, Wolffe, A. P, & Ge, H. (1997). Acetylation of general transcription factors by histone acetyltransferases. *Curr Biol* , 7, 689-692.

[57] Allfrey, V. G, & Mirsky, A. E. (1964). Structural Modifications of Histones and their Possible Role in the Regulation of RNA Synthesis. *Science* 144:559.

[58] Allfrey, V. G, Faulkner, R, & Mirsky, A. E. (1964). Acetylation and Methylation of Histones and Their Possible Role in the Regulation of Rna Synthesis. *Proc Natl Acad Sci U S A* , 51, 786-794.

[59] Brownell, J. E, Zhou, J, Ranalli, T, Kobayashi, R, Edmondson, D. G, Roth, S. Y, & Allis, C. D. (1996). Tetrahymena histone acetyltransferase A: a homolog to yeast Gcn5p linking histone acetylation to gene activation. *Cell* , 84, 843-851.

[60] Taunton, J, Hassig, C. A, & Schreiber, S. L. (1996). A mammalian histone deacetylase related to the yeast transcriptional regulator Rpd3p. *Science* , 272, 408-411.

[61] Dhalluin, C, Carlson, J. E, Zeng, L, He, C, Aggarwal, A. K, & Zhou, M. M. (1999). Structure and ligand of a histone acetyltransferase bromodomain. *Nature* , 399, 491-496.

[62] Rea, S, Eisenhaber, F, Carroll, O, Strahl, D, Sun, B. D, Schmid, Z. W, Opravil, M, Mechtler, S, Ponting, K, Allis, C. P, et al. (2000). Regulation of chromatin structure by site-specific histone H3 methyltransferases. *Nature* , 406, 593-599.

[63] Peters, A. H, Carroll, O, Scherthan, D, Mechtler, H, Sauer, K, Schofer, S, Weipoltshammer, C, Pagani, K, Lachner, M, Kohlmaier, M, et al. (2001). Loss of the Suv39h histone methyltransferases impairs mammalian heterochromatin and genome stability. *Cell* , 107, 323-337.

[64] Lachner, M, & Jenuwein, T. (2002). The many faces of histone lysine methylation. *Curr Opin Cell Biol* , 14, 286-298.

[65] Roh, T. Y, Cuddapah, S, Cui, K, & Zhao, K. (2006). The genomic landscape of histone modifications in human T cells. *Proc Natl Acad Sci U S A* , 103, 15782-15787.

[66] Guenther, M. G, Levine, S. S, Boyer, L. A, Jaenisch, R, & Young, R. A. (2007). A chromatin landmark and transcription initiation at most promoters in human cells. *Cell* , 130, 77-88.

[67] Jones, B. (2012). Chromatin: A model for nucleosome positioning. *Nat Rev Genet.*

[68] Jansen, A, Van Der Zande, E, Meert, W, Fink, G. R, & Verstrepen, K. J. (2012). Distal chromatin structure influences local nucleosome positions and gene expression. *Nucleic Acids Res* , 40, 3870-3885.

[69] Yen, K, Vinayachandran, V, Batta, K, Koerber, R. T, & Pugh, B. F. (2012). Genome-wide nucleosome specificity and directionality of chromatin remodelers. *Cell* , 149, 1461-1473.

[70] Tyler, J. K. (2002). Chromatin assembly. Cooperation between histone chaperones and ATP-dependent nucleosome remodeling machines. *Eur J Biochem* , 269, 2268-2274.

[71] Lewis, P. W, Elsaesser, S. J, Noh, K. M, Stadler, S. C, & Allis, C. D. (2010). Daxx is an H3.3-specific histone chaperone and cooperates with ATRX in replication-independent chromatin assembly at telomeres. *Proc Natl Acad Sci U S A* , 107, 14075-14080.

[72] Goldberg, A. D, Banaszynski, L. A, Noh, K. M, Lewis, P. W, Elsaesser, S. J, Stadler, S, Dewell, S, Law, M, Guo, X, Li, X, et al. (2010). Distinct factors control histone variant H3.3 localization at specific genomic regions. *Cell* , 140, 678-691.

[73] Gibbons, R. J, Picketts, D. J, Villard, L, & Higgs, D. R. (1995). Mutations in a putative global transcriptional regulator cause X-linked mental retardation with alpha-thalassemia (ATR-X syndrome). *Cell* , 80, 837-845.

[74] Gibbons, R. J, Picketts, D. J, & Higgs, D. R. (1995). Syndromal mental retardation due to mutations in a regulator of gene expression. *Hum Mol Genet* 4 Spec (1705-1709), 1705-1709.

[75] Luger, K, Dechassa, M. L, & Tremethick, D. J. (2012). New insights into nucleosome and chromatin structure: an ordered state or a disordered affair? *Nat Rev Mol Cell Biol* , 13, 436-447.

[76] Bartkuhn, M, & Renkawitz, R. (2008). Long range chromatin interactions involved in gene regulation. *Biochim Biophys Acta* , 1783, 2161-2166.

[77] Nolis, I. K, Mckay, D. J, Mantouvalou, E, Lomvardas, S, Merika, M, & Thanos, D. (2009). Transcription factors mediate long-range enhancer-promoter interactions. *Proc Natl Acad Sci U S A* , 106, 20222-20227.

[78] Bell, A. C, West, A. G, & Felsenfeld, G. (1999). The protein CTCF is required for the enhancer blocking activity of vertebrate insulators. *Cell*, 98, 387-396.

[79] Hou, C, Zhao, H, Tanimoto, K, & Dean, A. (2008). CTCF-dependent enhancer-blocking by alternative chromatin loop formation. *Proc Natl Acad Sci U S A*, 105, 20398-20403.

[80] Nativio, R, Wendt, K. S, Ito, Y, Huddleston, J. E, Uribe-lewis, S, Woodfine, K, Krueger, C, Reik, W, Peters, J. M, & Murrell, A. (2009). Cohesin is required for higher-order chromatin conformation at the imprinted IGFH19 locus. *PLoS Genet* 5:e1000739., 2.

[81] Phillips, J. E, & Corces, V. G. (2009). CTCF: master weaver of the genome. *Cell*, 137, 1194-1211.

[82] Botta, M, Haider, S, Leung, I. X, Lio, P, & Mozziconacci, J. (2010). Intra- and inter-chromosomal interactions correlate with CTCF binding genome wide. *Mol Syst Biol* 6:426.

[83] Wendt, K. S, Yoshida, K, Itoh, T, Bando, M, Koch, B, Schirghuber, E, Tsutsumi, S, Nagae, G, Ishihara, K, Mishiro, T, et al. (2008). Cohesin mediates transcriptional insulation by CCCTC-binding factor. *Nature*, 451, 796-801.

[84] Dorsett, D, & Krantz, I. D. (2009). On the molecular etiology of Cornelia de Lange syndrome. *Ann N Y Acad Sci*, 1151, 22-37.

[85] Huang, C. E, Milutinovich, M, & Koshland, D. (2005). Rings, bracelet or snaps: fashionable alternatives for Smc complexes. *Philos Trans R Soc Lond B Biol Sci*, 360, 537-542.

[86] Neuwald, A. F, & Hirano, T. (2000). HEAT repeats associated with condensins, cohesins, and other complexes involved in chromosome-related functions. *Genome Res*, 10, 1445-1452.

[87] Parelho, V, Hadjur, S, Spivakov, M, Leleu, M, Sauer, S, Gregson, H. C, Jarmuz, A, Canzonetta, C, Webster, Z, Nesterova, T, et al. (2008). Cohesins functionally associate with CTCF on mammalian chromosome arms. *Cell*, 132, 422-433.

[88] Xiao, T, Wallace, J, & Felsenfeld, G. (2011). Specific sites in the C terminus of CTCF interact with the SA2 subunit of the cohesin complex and are required for cohesin-dependent insulation activity. *Mol Cell Biol*, 31, 2174-2183.

[89] Guibert, S, Zhao, Z, Sjolinder, M, Gondor, A, Fernandez, A, Pant, V, & Ohlsson, R. (2012). CTCF-binding sites within the H19 ICR differentially regulate local chromatin structures and cis-acting functions. *Epigenetics*, 7, 361-369.

[90] Hadjur, S, Williams, L. M, Ryan, N. K, Cobb, B. S, Sexton, T, Fraser, P, Fisher, A. G, & Merkenschlager, M. (2009). Cohesins form chromosomal cis-interactions at the developmentally regulated IFNG locus. *Nature*, 460, 410-413.

[91] Chien, R, Zeng, W, Kawauchi, S, Bender, M. A, Santos, R, Gregson, H. C, Schmiesing, J. A, Newkirk, D. A, Kong, X, Ball, A. R, et al. (2011). Cohesin mediates chromatin

interactions that regulate mammalian beta-globin expression. *J Biol Chem* , 286, 17870-17878.

[92] Majumder, P, & Boss, J. M. (2011). Cohesin regulates MHC class II genes through interactions with MHC class II insulators. *J Immunol* , 187, 4236-4244.

[93] Ren, L, Shi, M, Wang, Y, Yang, Z, Wang, X, & Zhao, Z. (2012). CTCF and cohesin cooperatively mediate the cell-type specific interchromatin interaction between Bcl11b and Arhgap6 loci. *Mol Cell Biochem* , 360, 243-251.

[94] Ong, C. T, & Corces, V. G. (2011). Enhancer function: new insights into the regulation of tissue-specific gene expression. *Nat Rev Genet* , 12, 283-293.

[95] Kagey, M. H, Newman, J. J, Bilodeau, S, Zhan, Y, Orlando, D. A, Van Berkum, N. L, Ebmeier, C. C, Goossens, J, Rahl, P. B, Levine, S. S, et al. (2010). Mediator and cohesin connect gene expression and chromatin architecture. *Nature* , 467, 430-435.

[96] Oum, J. H, Seong, C, Kwon, Y, Ji, J. H, Sid, A, Ramakrishnan, S, Ira, G, Malkova, A, Sung, P, Lee, S. E, et al. (2011). RSC facilitates Rad59-dependent homologous recombination between sister chromatids by promoting cohesin loading at DNA double-strand breaks. *Mol Cell Biol* , 31, 3924-3937.

[97] Mcaleenan, A, Cordon-preciado, V, Clemente-blanco, A, Liu, I. C, Sen, N, Leonard, J, Jarmuz, A, & Aragon, L. (2012). SUMOylation of the alpha-Kleisin Subunit of Cohesin Is Required for DNA Damage-Induced Cohesion. *Curr Biol* , 22, 1564-1575.

[98] Caron, P, Aymard, F, Iacovoni, J. S, Briois, S, Canitrot, Y, Bugler, B, Massip, L, Losada, A, & Legube, G. (2012). Cohesin protects genes against gammaH2AX Induced by DNA double-strand breaks. *PLoS Genet* 8:e1002460.

[99] Nakanishi, M, Deardorff, M. A, Clark, D, Levy, S. E, Krantz, I, & Pipan, M. (2012). Investigation of autistic features among individuals with mild to moderate Cornelia de Lange syndrome. *Am J Med Genet A* 158A:, 1841-1847.

[100] Pearce, P. M, & Pitt, D. B. (1967). Six cases of de Lange's syndrome; parental consanguinity in two. *Med J Aust* , 1, 502-506.

[101] Opitz, J. M. (1985). The Brachmann-de Lange syndrome. *Am J Med Genet* , 22, 89-102.

[102] Noor, N, Kazmi, Z, & Mehnaz, A. (2012). Cornelia de Lange syndrome. *J Coll Physicians Surg Pak* , 22, 412-413.

[103] Verma, L, Passi, S, & Gauba, K. (2010). Brachman de Lange syndrome. *Contemp Clin Dent* , 1, 268-270.

[104] Kline, A. D, Krantz, I. D, Sommer, A, Kliewer, M, & Jackson, L. G. FitzPatrick, D.R., Levin, A.V., and Selicorni, A. (2007). Cornelia de Lange syndrome: clinical review, diagnostic and scoring systems, and anticipatory guidance. *Am J Med Genet A* 143A:, 1287-1296.

[105] Jackson, L, Kline, A. D, Barr, M. A, & Koch, S. (1993). de Lange syndrome: a clinical review of 310 individuals. *Am J Med Genet* , 47, 940-946.

[106] Mehta, A. V, & Ambalavanan, S. K. (1997). Occurrence of congenital heart disease in children with Brachmann-de Lange syndrome. *Am J Med Genet* , 71, 434-435.

[107] Chatfield, K. C, Schrier, S. A, Li, J, Clark, D, Kaur, M, Kline, A. D, Deardorff, M. A, Jackson, L. S, Goldmuntz, E, & Krantz, I. D. (2012). Congenital heart disease in Cornelia de Lange syndrome: Phenotype and genotype analysis. *Am J Med Genet A* 158A:, 2499-2505.

[108] Chong, K, Keating, S, Hurst, S, Summers, A, Berger, H, Seaward, G, Martin, N, Friedberg, T, & Chitayat, D. (2009). Cornelia de Lange syndrome (CdLS): prenatal and autopsy findings. *Prenat Diagn* , 29, 489-494.

[109] Cunniff, C, Curry, C. J, Carey, J. C, & Graham, J. M. Jr., Williams, C.A., Stengel-Rutkowski, S., Luttgen, S., and Meinecke, Congenital diaphragmatic hernia in the Brachmann-de Lange syndrome. *Am J Med Genet* 47:1018-1021., 1993.

[110] Mehta, G. D, Rizvi, S. M, & Ghosh, S. K. (2012). Cohesin: A guardian of genome integrity. *Biochim Biophys Acta* , 1823, 1324-1342.

[111] Kaur, M, Descipio, C, Mccallum, J, Yaeger, D, Devoto, M, Jackson, L. G, Spinner, N. B, & Krantz, I. D. (2005). Precocious sister chromatid separation (PSCS) in Cornelia de Lange syndrome. *Am J Med Genet A* , 138, 27-31.

[112] Muto, A, Calof, A. L, Lander, A. D, & Schilling, T. F. (2011). Multifactorial origins of heart and gut defects in nipbl-deficient zebrafish, a model of Cornelia de Lange Syndrome. *PLoS Biol* 9:e1001181.

[113] Kawauchi, S, Calof, A. L, Santos, R, Lopez-burks, M. E, Young, C. M, Hoang, M. P, Chua, A, Lao, T, Lechner, M. S, Daniel, J. A, et al. (2009). Multiple organ system defects and transcriptional dysregulation in the Nipbl(+/-) mouse, a model of Cornelia de Lange Syndrome. *PLoS Genet* 5:e1000650.

[114] Gillis, L. A, Mccallum, J, Kaur, M, Descipio, C, Yaeger, D, Mariani, A, Kline, A. D, Li, H. H, Devoto, M, Jackson, L. G, et al. (2004). NIPBL mutational analysis in 120 individuals with Cornelia de Lange syndrome and evaluation of genotype-phenotype correlations. *Am J Hum Genet* , 75, 610-623.

[115] Krantz, I. D, Mccallum, J, Descipio, C, Kaur, M, Gillis, L. A, Yaeger, D, Jukofsky, L, Wasserman, N, Bottani, A, Morris, C. A, et al. (2004). Cornelia de Lange syndrome is caused by mutations in NIPBL, the human homolog of Drosophila melanogaster Nipped-B. *Nat Genet* , 36, 631-635.

[116] Tonkin, E. T, Wang, T. J, Lisgo, S, Bamshad, M. J, & Strachan, T. (2004). NIPBL, encoding a homolog of fungal Scc2-type sister chromatid cohesion proteins and fly Nipped-B, is mutated in Cornelia de Lange syndrome. *Nat Genet* , 36, 636-641.

[117] Ciosk, R, Shirayama, M, Shevchenko, A, Tanaka, T, Toth, A, & Nasmyth, K. (2000). Cohesin's binding to chromosomes depends on a separate complex consisting of Scc2 and Scc4 proteins. *Mol Cell* , 5, 243-254.

[118] Schoumans, J, Wincent, J, Barbaro, M, Djureinovic, T, Maguire, P, Forsberg, L, Staaf, J, Thuresson, A. C, Borg, A, Nordgren, A, et al. (2007). Comprehensive mutational analysis of a cohort of Swedish Cornelia de Lange syndrome patients. *Eur J Hum Genet* , 15, 143-149.

[119] Pie, J, Gil-rodriguez, M. C, Ciero, M, Lopez-vinas, E, Ribate, M. P, Arnedo, M, Deardorff, M. A, Puisac, B, Legarreta, J, De Karam, J. C, et al. (2010). Mutations and variants in the cohesion factor genes NIPBL, SMC1A, and SMC3 in a cohort of 30 unrelated patients with Cornelia de Lange syndrome. *Am J Med Genet A* 152A:, 924-929.

[120] Deardorff, M. A, Bando, M, Nakato, R, Watrin, E, Itoh, T, Minamino, M, Saitoh, K, Komata, M, Katou, Y, Clark, D, et al. (2012). HDAC8 mutations in Cornelia de Lange syndrome affect the cohesin acetylation cycle. *Nature* , 489, 313-317.

[121] Harakalova, M. van den Boogaard, M.J., Sinke, R., van Lieshout, S., van Tuil, M.C., Duran, K., Renkens, I., Terhal, P.A., de Kovel, C., Nijman, I.J., et al. (2012). X-exome sequencing identifies a HDAC8 variant in a large pedigree with X-linked intellectual disability, truncal obesity, gynaecomastia, hypogonadism and unusual face. *J Med Genet* , 49, 539-543.

[122] Wu, Q, Zhang, T, Cheng, J. F, Kim, Y, Grimwood, J, Schmutz, J, Dickson, M, Noonan, J. P, Zhang, M. Q, Myers, R. M, et al. (2001). Comparative DNA sequence analysis of mouse and human protocadherin gene clusters. *Genome Res* , 11, 389-404.

[123] Hilschmann, N, Barnikol, H. U, Barnikol-watanabe, S, Gotz, H, Kratzin, H, & Thinnes, F. P. (2001). The immunoglobulin-like genetic predetermination of the brain: the protocadherins, blueprint of the neuronal network. *Naturwissenschaften* , 88, 2-12.

[124] Garrett, A. M, & Weiner, J. A. (2009). Control of CNS synapse development by {gamma}-protocadherin-mediated astrocyte-neuron contact. *J Neurosci* , 29, 11723-11731.

[125] Monahan, K, Rudnick, N. D, Kehayova, P. D, Pauli, F, Newberry, K. M, Myers, R. M, & Maniatis, T. Role of CCCTC binding factor (CTCF) and cohesin in the generation of single-cell diversity of protocadherin-alpha gene expression. *Proc Natl Acad Sci U S A* , 109, 9125-9130.

[126] Remeseiro, S, Cuadrado, A, Gomez-lopez, G, Pisano, D. G, & Losada, A. (2012). A unique role of cohesin-SA1 in gene regulation and development. *EMBO J* , 31, 2090-2102.

[127] Dorsett, D. (2004). Adherin: key to the cohesin ring and cornelia de Lange syndrome. *Curr Biol* 14:R, 834-836.

[128] Jahnke, P, Xu, W, Wulling, M, Albrecht, M, Gabriel, H, Gillessen-kaesbach, G, & Kaiser, F. J. (2008). The Cohesin loading factor NIPBL recruits histone deacetylases to mediate local chromatin modifications. *Nucleic Acids Res* , 36, 6450-6458.

[129] Liu, J, Zhang, Z, Bando, M, Itoh, T, Deardorff, M. A, Li, J. R, Clark, D, Kaur, M, Tatsuro, K, Kline, A. D, et al. (2010). Genome-wide DNA methylation analysis in cohesin mutant human cell lines. *Nucleic Acids Res* , 38, 5657-5671.

[130] Hagberg, B, Goutieres, F, Hanefeld, F, Rett, A, & Wilson, J. (1985). Rett syndrome: criteria for inclusion and exclusion. *Brain Dev* , 7, 372-373.

[131] Amir, R. E. Van den Veyver, I.B., Wan, M., Tran, C.Q., Francke, U., and Zoghbi, H.Y. (1999). Rett syndrome is caused by mutations in X-linked MECP2, encoding methyl-CpG-binding protein 2. *Nat Genet* , 23, 185-188.

[132] Meloni, I, Bruttini, M, Longo, I, Mari, F, Rizzolio, F, Adamo, D, Denvriendt, P, Fryns, K, Toniolo, J. P, & Renieri, D. A. (2000). A mutation in the rett syndrome gene, MECP2, causes X-linked mental retardation and progressive spasticity in males. *Am J Hum Genet* , 67, 982-985.

[133] Orrico, A, Lam, C, Galli, L, Dotti, M. T, Hayek, G, Tong, S. F, Poon, P. M, Zappella, M, Federico, A, & Sorrentino, V. (2000). MECP2 mutation in male patients with non-specific X-linked mental retardation. *FEBS Lett* , 481, 285-288.

[134] Van Esch, H, Bauters, M, Ignatius, J, Jansen, M, Raynaud, M, Hollanders, K, Lugtenberg, D, Bienvenu, T, Jensen, L. R, Gecz, J, et al. (2005). Duplication of the MECP2 region is a frequent cause of severe mental retardation and progressive neurological symptoms in males. *Am J Hum Genet* , 77, 442-453.

[135] Adkins, N.L, Georgel, P.T, & Me, . 2: structure and function. *Biochem Cell Biol* 89:1-11.

[136] Motil, K. J, Caeg, E, Barrish, J. O, Geerts, S, Lane, J. B, Percy, A. K, Annese, F, Mcnair, L, Skinner, S. A, Lee, H. S, et al. (2012). Gastrointestinal and nutritional problems occur frequently throughout life in girls and women with rett syndrome. *J Pediatr Gastroenterol Nutr* , 55, 292-298.

[137] Han, Z. A, Jeon, H. R, Kim, S. W, Park, J. Y, & Chung, H. J. (2012). Clinical characteristics of children with rett syndrome. *Ann Rehabil Med* , 36, 334-339.

[138] Nan, X, Ng, H. H, Johnson, C. A, Laherty, C. D, Turner, B. M, Eisenman, R. N, & Bird, A. (1998). Transcriptional repression by the methyl-CpG-binding protein MeCP2 involves a histone deacetylase complex. *Nature* , 393, 386-389.

[139] Balaratnasingam, S, & Janca, A. (2012). Brain Derived Neurotrophic Factor: a novel neurotrophin involved in psychiatric and neurological disorders. *Pharmacol Ther* , 134, 116-124.

[140] Chen, W. G, Chang, Q, Lin, Y, Meissner, A, West, A. E, Griffith, E. C, Jaenisch, R, & Greenberg, M. E. (2003). Derepression of BDNF transcription involves calcium-dependent phosphorylation of MeCP2. *Science* , 302, 885-889.

[141] Martinowich, K, Hattori, D, Wu, H, Fouse, S, He, F, Hu, Y, Fan, G, & Sun, Y. E. (2003). DNA methylation-related chromatin remodeling in activity-dependent BDNF gene regulation. *Science* , 302, 890-893.

[142] Chahrour, M, Jung, S.Y, Shaw, C, Zhou, X, Wong, S.T, Qin, J, Zoghbi, H.Y, & Me, . 2, a key contributor to neurological disease, activates and represses transcription. *Science* 320:1224-1229.

[143] Skene, P. J, Illingworth, R. S, Webb, S, Kerr, A. R, James, K. D, Turner, D. J, Andrews, R, & Bird, A. P. (2010). Neuronal MeCP2 is expressed at near histone-octamer levels and globally alters the chromatin state. *Mol Cell* , 37, 457-468.

[144] Ghosh, R.P, Horowitz-Scherer, R.A, Nikitina, T, Shlyakhtenko, L.S, Woodcock, C.L, & Me, . 2 binds cooperatively to its substrate and competes with histone H1 for chromatin binding sites. *Mol Cell Biol* 30:4656-4670.

[145] Weitzel, J. M, Buhrmester, H, & Stratling, W. H. (1997). Chicken MAR-binding protein ARBP is homologous to rat methyl-CpG-binding protein MeCP2. *Mol Cell Biol* , 17, 5656-5666.

[146] Buhrmester, H, Von Kries, J. P, & Stratling, W. H. (1995). Nuclear matrix protein ARBP recognizes a novel DNA sequence motif with high affinity. *Biochemistry* , 34, 4108-4117.

[147] Horike, S, Cai, S, Miyano, M, Cheng, J. F, & Kohwi-shigematsu, T. (2005). Loss of silent-chromatin looping and impaired imprinting of DLX5 in Rett syndrome. *Nat Genet* , 37, 31-40.

[148] Wang, Y, Dye, C. A, Sohal, V, Long, J. E, Estrada, R. C, Roztocil, T, Lufkin, T, Deisseroth, K, Baraban, S. C, & Rubenstein, J. L. (2010). Dlx5 and Dlx6 regulate the development of parvalbumin-expressing cortical interneurons. *J Neurosci* , 30, 5334-5345.

[149] Hernandez-miranda, L. R, Parnavelas, J. G, & Chiara, F. Molecules and mechanisms involved in the generation and migration of cortical interneurons. *ASN Neuro* 2:e00031.

[150] Chao, H. T, Chen, H, Samaco, R. C, Xue, M, Chahrour, M, Yoo, J, Neul, J. L, Gong, S, Lu, H. C, Heintz, N, et al. (2010). Dysfunction in GABA signalling mediates autism-like stereotypies and Rett syndrome phenotypes. *Nature* , 468, 263-269.

[151] Brero, A, Easwaran, H. P, Nowak, D, Grunewald, I, Cremer, T, Leonhardt, H, & Cardoso, M. C. (2005). Methyl CpG-binding proteins induce large-scale chromatin reorganization during terminal differentiation. *J Cell Biol* , 169, 733-743.

[152] Agarwal, N, Becker, A, Jost, K.L, Haase, S, Thakur, B.K, Brero, A, Hardt, T, Kudo, S, Leonhardt, H, Cardoso, M.C, & Me, . 2 Rett mutations affect large scale chromatin organization. *Hum Mol Genet* 20:4187-4195.

[153] Nikitina, T, Ghosh, R.P, Horowitz-Scherer, R.A, Hansen, J.C, Grigoryev, S.A, Woodcock, C.L, & Me, . 2-chromatin interactions include the formation of chromatosome-like structures and are altered in mutations causing Rett syndrome. *J Biol Chem* 282:28237-28245.

[154] Georgel, P. T, Horowitz-scherer, R. A, Adkins, N, Woodcock, C. L, Wade, P. A, & Hansen, J. C. (2003). Chromatin compaction by human MeCP2. Assembly of novel secondary chromatin structures in the absence of DNA methylation. *J Biol Chem* , 278, 32181-32188.

[155] Kernohan, K. D, Jiang, Y, Tremblay, D. C, Bonvissuto, A. C, Eubanks, J. H, Mann, M. R, & Berube, N. G. (2010). ATRX partners with cohesin and MeCP2 and contributes to developmental silencing of imprinted genes in the brain. *Dev Cell* , 18, 191-202.

[156] Casas-delucchi, C. S, Becker, A, Bolius, J. J, & Cardoso, M. C. (2012). Targeted manipulation of heterochromatin rescues MeCP2 Rett mutants and re-establishes higher order chromatin organization. *Nucleic Acids Res.*

[157] Calfa, G, Percy, A. K, & Pozzo-miller, L. (2011). Experimental models of Rett syndrome based on Mecp2 dysfunction. *Exp Biol Med (Maywood)* , 236, 3-19.

[158] Chen, R. Z, Akbarian, S, Tudor, M, & Jaenisch, R. (2001). Deficiency of methyl-CpG binding protein-2 in CNS neurons results in a Rett-like phenotype in mice. *Nat Genet* , 27, 327-331.

[159] Guy, J, Hendrich, B, Holmes, M, Martin, J. E, & Bird, A. (2001). A mouse Mecp2-null mutation causes neurological symptoms that mimic Rett syndrome. *Nat Genet* , 27, 322-326.

[160] Collins, A. L, Levenson, J. M, Vilaythong, A. P, Richman, R, Armstrong, D. L, & Noebels, J. L. David Sweatt, J., and Zoghbi, H.Y. (2004). Mild overexpression of MeCP2 causes a progressive neurological disorder in mice. *Hum Mol Genet* , 13, 2679-2689.

[161] Cochran, K. W, & Allen, L. B. (1970). Simple method of evaluating scrapie in mice. *Appl Microbiol* , 20, 72-74.

[162] Nguyen, M.V, Du, F, Felice, C.A, Shan, X, Nigam, A, Mandel, G, Robinson, J.K, Ballas, N, & Me, . 2 is critical for maintaining mature neuronal networks and global brain anatomy during late stages of postnatal brain development and in the mature adult brain. *J Neurosci* 32:10021-10034.

[163] Weng, S. M, Bailey, M. E, & Cobb, S. R. (2011). Rett syndrome: from bed to bench. *Pediatr Neonatol* , 52, 309-316.

[164] Asaka, Y, Jugloff, D. G, Zhang, L, Eubanks, J. H, & Fitzsimonds, R. M. (2006). Hippocampal synaptic plasticity is impaired in the Mecp2-null mouse model of Rett syndrome. *Neurobiol Dis* , 21, 217-227.

[165] Moretti, P, Levenson, J. M, Battaglia, F, Atkinson, R, Teague, R, Antalffy, B, Armstrong, D, Arancio, O, Sweatt, J. D, & Zoghbi, H. Y. (2006). Learning and memory and synaptic plasticity are impaired in a mouse model of Rett syndrome. *J Neurosci* , 26, 319-327.

[166] Dani, V. S, Chang, Q, Maffei, A, Turrigiano, G. G, Jaenisch, R, & Nelson, S. B. (2005). Reduced cortical activity due to a shift in the balance between excitation and inhibition in a mouse model of Rett syndrome. *Proc Natl Acad Sci U S A* , 102, 12560-12565.

[167] Nelson, E.D, Kavalali, E.T, Monteggia, L.M, & Me, . 2-dependent transcriptional repression regulates excitatory neurotransmission. *Curr Biol* 16:710-716.

[168] Stuss, D.P, Boyd, J.D, Levin, D.B, Delaney, K.R, & Me, . 2 mutation results in compartment-specific reductions in dendritic branching and spine density in layer 5 motor cortical neurons of YFP-H mice. *PLoS One* 7:e31896.

[169] Belichenko, P. V, Wright, E. E, Belichenko, N. P, Masliah, E, Li, H. H, Mobley, W. C, & Francke, U. (2009). Widespread changes in dendritic and axonal morphology in Mecp2-mutant mouse models of Rett syndrome: evidence for disruption of neuronal networks. *J Comp Neurol* , 514, 240-258.

[170] Weatherall, D. J, Higgs, D. R, Bunch, C, Old, J. M, Hunt, D. M, Pressley, L, Clegg, J. B, Bethlenfalvay, N. C, Sjolin, S, Koler, R. D, et al. (1981). Hemoglobin H disease and mental retardation: a new syndrome or a remarkable coincidence? *N Engl J Med* , 305, 607-612.

[171] Gibbons, R. J, Wilkie, A. O, Weatherall, D. J, & Higgs, D. R. (1991). A newly defined X linked mental retardation syndrome associated with alpha thalassaemia. *J Med Genet* , 28, 729-733.

[172] Medina, C. F, Mazerolle, C, Wang, Y, Berube, N. G, Coupland, S, Gibbons, R. J, Wallace, V. A, & Picketts, D. J. (2009). Altered visual function and interneuron survival in Atrx knockout mice: inference for the human syndrome. *Hum Mol Genet* , 18, 966-977.

[173] Gibbons, R. J, Suthers, G. K, Wilkie, A. O, Buckle, V. J, & Higgs, D. R. mental retardation (ATR-X) syndrome: localization to Xq12-q21.31 by X inactivation and linkage analysis. *Am J Hum Genet* , 51, 1136-1149.

[174] Wada, T, Sugie, H, Fukushima, Y, & Saitoh, S. (2005). Non-skewed X-inactivation may cause mental retardation in a female carrier of X-linked alpha-thalassemia/mental retardation syndrome (ATR-X): X-inactivation study of nine female carriers of ATR-X. *Am J Med Genet A* , 138, 18-20.

[175] Gibbons, R. J, & Higgs, D. R. (2000). Molecular-clinical spectrum of the ATR-X syndrome. *Am J Med Genet* , 97, 204-212.

[176] Villard, L, Fontes, M, Ades, L. C, & Gecz, J. (2000). Identification of a mutation in the XNP/ATR-X gene in a family reported as Smith-Fineman-Myers syndrome. *Am J Med Genet* , 91, 83-85.

[177] Gibbons, R. (2006). Alpha thalassaemia-mental retardation, X linked. *Orphanet J Rare Dis* 1:15.

[178] Martucciello, G, Lombardi, L, Savasta, S, & Gibbons, R. J. (2006). Gastrointestinal phenotype of ATR-X syndrome. *Am J Med Genet A* , 140, 1172-1176.

[179] Villard, L, Gecz, J, Mattei, J. F, Fontes, M, Saugier-veber, P, Munnich, A, & Lyonnet, S. (1996). XNP mutation in a large family with Juberg-Marsidi syndrome. *Nat Genet* , 12, 359-360.

[180] Abidi, F, Schwartz, C. E, Carpenter, N. J, Villard, L, Fontes, M, & Curtis, M. (1999). Carpenter-Waziri syndrome results from a mutation in XNP. *Am J Med Genet* , 85, 249-251.

[181] Martinez, F, Tomas, M, Millan, J. M, Fernandez, A, Palau, F, & Prieto, F. (1998). Genetic localisation of mental retardation with spastic diplegia to the pericentromeric region of the X chromosome: X inactivation in female carriers. *J Med Genet* , 35, 284-287.

[182] Gibbons, R. J, Wada, T, Fisher, C. A, Malik, N, Mitson, M. J, Steensma, D. P, Fryer, A, Goudie, D. R, Krantz, I. D, & Traeger-synodinos, J. (2008). Mutations in the chromatin-associated protein ATRX. *Hum Mutat* , 29, 796-802.

[183] Cardoso, C, Lutz, Y, Mignon, C, Compe, E, Depetris, D, Mattei, M. G, Fontes, M, & Colleaux, L. (2000). ATR-X mutations cause impaired nuclear location and altered DNA binding properties of the XNP/ATR-X protein. *J Med Genet* , 37, 746-751.

[184] Picketts, D. J, Tastan, A. O, Higgs, D. R, & Gibbons, R. J. (1998). Comparison of the human and murine ATRX gene identifies highly conserved, functionally important domains. *Mamm Genome* , 9, 400-403.

[185] Argentaro, A, Yang, J. C, Chapman, L, Kowalczyk, M. S, Gibbons, R. J, Higgs, D. R, Neuhaus, D, & Rhodes, D. (2007). Structural consequences of disease-causing mutations in the ATRX-DNMT3-DNMT3L (ADD) domain of the chromatin-associated protein ATRX. *Proc Natl Acad Sci U S A* , 104, 11939-11944.

[186] Iwase, S, Xiang, B, Ghosh, S, Ren, T, Lewis, P. W, Cochrane, J. C, Allis, C. D, Picketts, D. J, Patel, D. J, Li, H, et al. (2011). ATRX ADD domain links an atypical histone methylation recognition mechanism to human mental-retardation syndrome. *Nat Struct Mol Biol* , 18, 769-776.

[187] Dhayalan, A, Tamas, R, Bock, I, Tattermusch, A, Dimitrova, E, Kudithipudi, S, Ragozin, S, & Jeltsch, A. (2011). The ATRX-ADD domain binds to H3 tail peptides and reads the combined methylation state of K4 and K9. *Hum Mol Genet* , 20, 2195-2203.

[188] Richmond, E, & Peterson, C. L. (1996). Functional analysis of the DNA-stimulated ATPase domain of yeast SWI2/SNF2. *Nucleic Acids Res* , 24, 3685-3692.

[189] Berube, N. G, Smeenk, C. A, & Picketts, D. J. (2000). Cell cycle-dependent phosphorylation of the ATRX protein correlates with changes in nuclear matrix and chromatin association. *Hum Mol Genet* , 9, 539-547.

[190] Lechner, M. S, Schultz, D. C, Negorev, D, Maul, G. G, & Rauscher, F. J. rd. (2005). The mammalian heterochromatin protein 1 binds diverse nuclear proteins through a common motif that targets the chromoshadow domain. *Biochem Biophys Res Commun* , 331, 929-937.

[191] Cardoso, C, Timsit, S, Villard, L, Khrestchatisky, M, Fontes, M, & Colleaux, L. (1998). Specific interaction between the XNP/ATR-X gene product and the SET domain of the human EZH2 protein. *Hum Mol Genet* , 7, 679-684.

[192] Nan, X, Hou, J, Maclean, A, Nasir, J, Lafuente, M. J, Shu, X, Kriaucionis, S, & Bird, A. (2007). Interaction between chromatin proteins MECP2 and ATRX is disrupted by mutations that cause inherited mental retardation. *Proc Natl Acad Sci U S A* , 104, 2709-2714.

[193] Tang, J, Wu, S, Liu, H, Stratt, R, Barak, O. G, Shiekhattar, R, Picketts, D. J, & Yang, X. (2004). A novel transcription regulatory complex containing death domain-associated protein and the ATR-X syndrome protein. *J Biol Chem* , 279, 20369-20377.

[194] Xue, Y, Gibbons, R, Yan, Z, Yang, D, Mcdowell, T. L, Sechi, S, Qin, J, Zhou, S, Higgs, D, & Wang, W. (2003). The ATRX syndrome protein forms a chromatin-remodeling complex with Daxx and localizes in promyelocytic leukemia nuclear bodies. *Proc Natl Acad Sci U S A* , 100, 10635-10640.

[195] Law, M. J, Lower, K. M, Voon, H. P, Hughes, J. R, Garrick, D, Viprakasit, V, Mitson, M, De Gobbi, M, Marra, M, Morris, A, et al. (2010). ATR-X syndrome protein targets tandem repeats and influences allele-specific expression in a size-dependent manner. *Cell* , 143, 367-378.

[196] Mcdowell, T. L, Gibbons, R. J, Sutherland, H, Rourke, O, Bickmore, D. M, Pombo, W. A, Turley, A, Gatter, H, Picketts, K, Buckle, D. J, et al. (1999). Localization of a putative transcriptional regulator (ATRX) at pericentromeric heterochromatin and the short arms of acrocentric chromosomes. *Proc Natl Acad Sci U S A* , 96, 13983-13988.

[197] Berube, N. G. (2011). ATRX in chromatin assembly and genome architecture during development and disease. *Biochem Cell Biol* , 89, 435-444.

[198] Kourmouli, N, Sun, Y. M, Van Der Sar, S, Singh, P. B, & Brown, J. P. (2005). Epigenetic regulation of mammalian pericentric heterochromatin in vivo by HP1. *Biochem Biophys Res Commun* , 337, 901-907.

[199] Berube, N. G, Mangelsdorf, M, Jagla, M, Vanderluit, J, Garrick, D, Gibbons, R. J, Higgs, D. R, Slack, R. S, & Picketts, D. J. (2005). The chromatin-remodeling protein

ATRX is critical for neuronal survival during corticogenesis. *J Clin Invest* , 115, 258-267.

[200] Seah, C, Levy, M. A, Jiang, Y, Mokhtarzada, S, Higgs, D. R, Gibbons, R. J, & Berube, N. G. (2008). Neuronal death resulting from targeted disruption of the Snf2 protein ATRX is mediated by *J Neurosci* 28:12570 12580., 53.

[201] Nogami, T, Beppu, H, Tokoro, T, Moriguchi, S, Shioda, N, Fukunaga, K, Ohtsuka, T, Ishii, Y, Sasahara, M, Shimada, Y, et al. (2011). Reduced expression of the ATRX gene, a chromatin-remodeling factor, causes hippocampal dysfunction in mice. *Hippocampus* , 21, 678-687.

[202] Shioda, N, Beppu, H, Fukuda, T, Li, E, Kitajima, I, & Fukunaga, K. (2011). Aberrant calcium/calmodulin-dependent protein kinase II (CaMKII) activity is associated with abnormal dendritic spine morphology in the ATRX mutant mouse brain. *J Neurosci* , 31, 346-358.

[203] Levy, M. A, Fernandes, A. D, Tremblay, D. C, Seah, C, & Berube, N. G. (2008). The SWI/SNF protein ATRX co-regulates pseudoautosomal genes that have translocated to autosomes in the mouse genome. *BMC Genomics* 9:468.

[204] Hatton, C. S, Wilkie, A. O, Drysdale, H. C, Wood, W. G, Vickers, M. A, Sharpe, J, Ayyub, H, Pretorius, I. M, Buckle, V. J, & Higgs, D. R. (1990). Alpha-thalassemia caused by a large (62 kb) deletion upstream of the human alpha globin gene cluster. *Blood* , 76, 221-227.

[205] Ratnakumar, K, & Duarte, L. F. LeRoy, G., Hasson, D., Smeets, D., Vardabasso, C., Bonisch, C., Zeng, T., Xiang, B., Zhang, D.Y., et al. (2012). ATRX-mediated chromatin association of histone variant macroH2A1 regulates alpha-globin expression. *Genes Dev* , 26, 433-438.

[206] Drane, P, Ouararhni, K, Depaux, A, Shuaib, M, & Hamiche, A. (2010). The death-associated protein DAXX is a novel histone chaperone involved in the replication-independent deposition of H3.3. *Genes Dev* , 24, 1253-1265.

[207] Eustermann, S, Yang, J. C, Law, M. J, Amos, R, Chapman, L. M, Jelinska, C, Garrick, D, Clynes, D, Gibbons, R. J, Rhodes, D, et al. (2011). Combinatorial readout of histone H3 modifications specifies localization of ATRX to heterochromatin. *Nat Struct Mol Biol* , 18, 777-782.

[208] Hashimoto, H, Vertino, P. M, & Cheng, X. (2010). Molecular coupling of DNA methylation and histone methylation. *Epigenomics* , 2, 657-669.

[209] Honda, S, Satomura, S, Hayashi, S, Imoto, I, Nakagawa, E, Goto, Y, & Inazawa, J. and Japanese Mental Retardation, C. (2012). Concomitant microduplications of MECP2 and ATRX in male patients with severe mental retardation. *J Hum Genet* , 57, 73-77.

[210] Ritchie, K, Seah, C, Moulin, J, Isaac, C, Dick, F, & Berube, N. G. (2008). Loss of ATRX leads to chromosome cohesion and congression defects. *J Cell Biol* , 180, 315-324.

[211] Armstrong, D, Dunn, J. K, Antalffy, B, & Trivedi, R. (1995). Selective dendritic alterations in the cortex of Rett syndrome. *J Neuropathol Exp Neurol* , 54, 195-201.

[212] Gadalla, K.K, Bailey, M.E, Cobb, S.R, & Me, . 2 and Rett syndrome: reversibility and potential avenues for therapy. *Biochem J* 439:1-14.

[213] Guy, J, Gan, J, Selfridge, J, Cobb, S, & Bird, A. (2007). Reversal of neurological defects in a mouse model of Rett syndrome. *Science* , 315, 1143-1147.

[214] Luikenhuis, S, Giacometti, E, Beard, C. F, & Jaenisch, R. (2004). Expression of MeCP2 in postmitotic neurons rescues Rett syndrome in mice. *Proc Natl Acad Sci U S A* , 101, 6033-6038.

[215] Robinson, L, Guy, J, Mckay, L, Brockett, E, Spike, R. C, Selfridge, J, De Sousa, D, Merusi, C, Riedel, G, Bird, A, et al. (2012). Morphological and functional reversal of phenotypes in a mouse model of Rett syndrome. *Brain* , 135, 2699-2710.

[216] Cobb, S, Guy, J, & Bird, A. (2010). Reversibility of functional deficits in experimental models of Rett syndrome. *Biochem Soc Trans* , 38, 498-506.

[217] Gray, S. J. (2012). Gene therapy and neurodevelopmental disorders. *Neuropharmacology*.

[218] Friez, M. J, Jones, J. R, Clarkson, K, Lubs, H, Abuelo, D, Bier, J. A, Pai, S, Simensen, R, Williams, C, Giampietro, P. F, et al. (2006). Recurrent infections, hypotonia, and mental retardation caused by duplication of MECP2 and adjacent region in Xq28. *Pediatrics* 118:e, 1687-1695.

[219] Meins, M, Lehmann, J, Gerresheim, F, Herchenbach, J, Hagedorn, M, Hameister, K, & Epplen, J. T. (2005). Submicroscopic duplication in Xq28 causes increased expression of the MECP2 gene in a boy with severe mental retardation and features of Rett syndrome. *J Med Genet* 42:e12.

[220] Berube, N. G, Jagla, M, Smeenk, C, De Repentigny, Y, Kothary, R, & Picketts, D. J. (2002). Neurodevelopmental defects resulting from ATRX overexpression in transgenic mice. *Hum Mol Genet* , 11, 253-261.

[221] Lugtenberg, D, De Brouwer, A. P, Oudakker, A. R, Pfundt, R, Hamel, B. C, Van Bokhoven, H, & Bongers, E. M. duplication encompassing the ATRX gene in a man with mental retardation, minor facial and genital anomalies, short stature and broad thorax. *Am J Med Genet A* 149A:, 760-766.

[222] Cohn, D. M, Pagon, R. A, Hudgins, L, Schwartz, C. E, Stevenson, R. E, & Friez, M. J. (2009). Partial ATRX gene duplication causes ATR-X syndrome. *Am J Med Genet A* 149A:, 2317-2320.

[223] Tsai, S. J. (2012). Peripheral administration of brain-derived neurotrophic factor to Rett syndrome animal model: A possible approach for the treatment of Rett syndrome. *Med Sci Monit* 18:HY, 33-36.

[224] Chang, Q, Khare, G, Dani, V, Nelson, S, & Jaenisch, R. (2006). The disease progression of Mecp2 mutant mice is affected by the level of BDNF expression. *Neuron* , 49, 341-348.

[225] Schmid, D. A, Yang, T, Ogier, M, Adams, I, Mirakhur, Y, Wang, Q, Massa, S. M, Longo, F. M, & Katz, D. M. B small molecule partial agonist rescues TrkB phosphorylation deficits and improves respiratory function in a mouse model of Rett syndrome. *J Neurosci* , 32, 1803-1810.

[226] Balkowiec, A, Kunze, D. L, & Katz, D. M. (2000). Brain-derived neurotrophic factor acutely inhibits AMPA-mediated currents in developing sensory relay neurons. *J Neurosci* , 20, 1904-1911.

[227] Kline, D. D, Ogier, M, Kunze, D. L, & Katz, D. M. (2010). Exogenous brain-derived neurotrophic factor rescues synaptic dysfunction in Mecp2-null mice. *J Neurosci* , 30, 5303-5310.

[228] Kron, M, Reuter, J, Gerhardt, E, Manzke, T, Zhang, W, & Dutschmann, M. (2008). Emergence of brain-derived neurotrophic factor-induced postsynaptic potentiation of NMDA currents during the postnatal maturation of the Kolliker-Fuse nucleus of rat. *J Physiol* , 586, 2331-2343.

[229] Bondy, C. A, & Cheng, C. M. (2004). Signaling by insulin-like growth factor 1 in brain. *Eur J Pharmacol* , 490, 25-31.

[230] Tropea, D, Giacometti, E, Wilson, N. R, Beard, C, Mccurry, C, Fu, D. D, Flannery, R, Jaenisch, R, & Sur, M. (2009). Partial reversal of Rett Syndrome-like symptoms in MeCP2 mutant mice. *Proc Natl Acad Sci U S A* , 106, 2029-2034.

[231] Chailangkarn, T, Acab, A, & Muotri, A. R. (2012). Modeling neurodevelopmental disorders using human neurons. *Curr Opin Neurobiol.*

[232] Takahashi, A, Tokunaga, A, Yamanaka, H, Mashimo, T, Noguchi, K, & Uchida, I. (2006). Two types of GABAergic miniature inhibitory postsynaptic currents in mouse substantia gelatinosa neurons. *Eur J Pharmacol* , 553, 120-128.

[233] Takahashi, K, Okita, K, Nakagawa, M, & Yamanaka, S. (2007). Induction of pluripotent stem cells from fibroblast cultures. *Nat Protoc* , 2, 3081-3089.

[234] Marchetto, M. C, Carromeu, C, Acab, A, Yu, D, Yeo, G. W, Mu, Y, Chen, G, Gage, F. H, & Muotri, A. R. (2010). A model for neural development and treatment of Rett syndrome using human induced pluripotent stem cells. *Cell* , 143, 527-539.

Chapter 3

Intellectual and Behavioral Disabilities in Smith – Magenis Syndrome

Danilo Moretti-Ferreira

Additional information is available at the end of the chapter

http://dx.doi.org/10.5772/55721

1. Introduction

Smith-Magenis syndrome (SMS) is a rare developmental disorder featuring impaired intellectual and behavioral abnormalities. SMS is still not well known because it is characterized by subtle facial dysmorphology that progresses with age, and clinical features that overlap with other intellectual disability syndromes as Prader–Willi, Williams-Beuren, and Down syndromes. Due to their intellectual impairment especially their abnormal and frequently anti-social behavior, most individuals affected with SMS are institutionalized without proper diagnosis and care.

2. Background

Patients with the features of SMS were first described in 1982 by Ann C.M. Smith in an abstract presented at the Annual Meeting of the American Society of Human Genetics [1]. In 1986, Ellen Magenis together with Ann C.M. Smith and their colleagues published a clinical review of nine individuals affected by this nosologic entity. For this reason the syndrome was named after them [2]. A deletion of chromosome 17p11.2 was identified as the cause of this condition in approximately 90% of cases, and thus this disorder belongs to the group of contiguous-gene syndromes, currently referred to as genome diseases [3-5]. SMS patients without deletions 17p11.2 may carry a point mutation in the gene *RAI1* [6-10], which codes a transcription factor acting in several different biological pathways. *RAI1* dosage is crucial for normal regulation of circadian rhythm, lipid metabolism, and melatonin function. SMS affects both sexes equally and has been found in all ethnic groups. The incidence of SMS was initially estimated at 1:25,000 births [11]. However improvements in cytogenetic techniques and molecular analyses have allowed the diagnosis of most cases, leading to a current prevalence estimate of 1:15,000 [12].

3. Clinical characteristics

SMS dysmorphysms change with age. The most common facial characteristics of the syndrome include broad square-shaped face, brachycephaly, prominent forehead, synophrys, deep-set eyes, broad nasal bridge, midface hypoplasia, micrognathia in infancy, relative prognathism with age, and everted, "tented" upper lip [13]. Dental anomalies such as premolar agenesis and taurodontism have also been reported [14].

Individuals with SMS show mild to moderate mental retardation [11, 15], and behavioral abnormalities such as sleep disturbances, stereotypic movement, and self-injurious behavior [16-21].

To date, nearly two hundred cases have been described in the literature. In 2011 Gamba and colleagues presented seven Brazilian cases and a meta-analysis of clinical signs in SMS reported in the literature, which are summarized in the table below [22].

Clinical	Gamba et al., 2011 N = 7	Literature N = 165		Fisher's Exact test
Craniofacial			%	*P value*
Brachycephaly	5/7	95/106	89.6	0.1893
Microcephaly	3/7	9/56	16.0	0.1199
Midface Hypoplasia	6/7	87/10	79.8	1.0000
Broad, square-shaped face	7/7	64/82	78.0	0.3367
Broad Nasal Bridge	7/7	41/51	80.39	0.3356
Short Philtrum	7/7	11/11	100.00	1.0000
Everted, "tented" upper lip	6/7	64/83	77.11	1.0000
Cleft lip/palate	0/7	12/47	25.53	0.3275
Relative prognathism with age	6/7	49/62	79.03	1.0000
Micrognathia	1/7	12/28	42.86	0.2197
Skeletal				
Short stature	6/7	35/71	49.30	0.1115
Scoliosis	3/7	23/53	43.40	0.6971
Dental anomalies	7/7	4/11	36.36	0.0128
Short broad hands	7/7	n/a	n/a	
Clinodactyly	6/7	19/30	63.33	0.3891
Brachydactyly	7/7	67/81	82.72	0.5915
Syndactly	5/7	15/50	30.00	0.0837
Ocular abnormalities				
Deep-set, close-spaced eyes	6/7	47/72	65.28	0.4156
Synophrys	5/7	31/57	54.39	0.4540

Clinical	Gamba et al., 2011 N = 7	Literature N = 165		Fisher's Exact test
Strabismus	7/7	39/67	58.21	0.0400
Iris Abnormalities	5/7	10/23	43.48	0.3898
Otoryngological				
Ear Abnormalities	4/6	48/76	63.16	0.6938
Ear infections	2/7	28/36	77.78	0.0190
Hoarse, deep voice	1/7	40/52	76.92	0.0023
Hearing loss	3/7	46/74	62.16	0.4258
Neurological				
Cognitive impairment/developmental delay	7/7	100/100	100.00	1.0000
Speech delay	7/7	101/111	90.99	1.0000
Motor Delay	7/7	92/114	80.70	0.3479
Infantile hypotonia	4/7	49/77	63.64	0.7054
Sleep disturbance	5/6	97/110	88.18	0.5462
Hyporeflexia	1/7	2/4	50.00	0.4909
Behavior				
Self-Hung	7/7	17/20	85.00	1.0000
Onychotillomania	5/7	15/24	62.50	1.0000
Polyembolokoilamania	6/7	14/23	60.87	0.3717
Head Banging/Face Slapping	1/2	36/43	83.72	0.3273
Hand Biting	4/5	19/19	100.00	0.2083
Attention Seeking	5/5	52/54	96.30	1.0000
Aggressive behavior	7/7	62/67	92.54	1.0000
Self-injurious behaviors	7/7	56/61	91.80	1.0000
Hyperactivity	6/7	52/54	96.30	1.0000
Other features				
Cardiac defects	0/7	44/88	50.00	0.0139
Renal/urinary tract abnormalities	1/7	12/49	24.49	1.0000
EEG abnormal/ evident seizures	3/5	23/58	39.66	0.6687
Hypogonadism male	1/7	21/60	35.00	0.4116
Obesity	3/7	12/51	23.53	0.3597

Bold type Statistically significant values of P less than 0.05 (two-tailed Fisher exact test).

(+) positive (-) negative (n/a) not available

Table 1. Clinical features of seven Brazilian Smith-Magenis syndrome cases and meta-analysis of 165 cases from the literature (Gamba et al., Genet. Mol. Res. 10 [4]: 2664-2670, 2011, Published with permission from *Genetics and Molecular Research-Online Journal*) [22].

3.1. SMS features in infancy, childhood/adolescence and adulthood

3.1.1. Infancy

The gestation of children with SMS is commonly uneventful. When maternal report is available, a diminution of fetal movements is described in 50% of cases. At birth all parameters (weight, length, OFC) are normal, including time of gestation.

The occurrence of generalized hypotonia and hyporeflexia promotes a marked oral motor dysfunction, with poor sucking and swallowing, and gastroesophageal reflux. Failure to thrive is attributed to feeding difficulties. During the first year of life, parents often describe SMS cases as perfect babies because they sleep very well and cry little.

Behavior disturbances can be observed as early as 4 months. Videotape analysis shows patients' motor repertoire is significantly reduced, and fidgety general movements, which are typical of that age, are missing. Posture is abnormal and overall movements are jerky and monotonous. These findings indicate a severe motor impairment as early as 4 months of age [23]. Beyond 18 months, signs of developmental delay become increasingly obvious, with early stages of intense crying and sleepless nights. Within 2-3 years of age patients have a clear delay in language acquisition, with lalation [24-25]. Dysmorphic signs subsequently begin to become more evident, with facial hypotonia, and relative micrognathia.

3.1.2. Childhood/adolescence

It is at this stage of life that patients with SBS have dysmorphisms, significant cognitive delays and a peculiar behavior and come to the attention of health professionals. Most patients are diagnosed at this stage of life.

Facial dysmorphisms include broad and square-shaped face, mild face hypoplasia, brachicephaly, short nasal philtrum, a tendency toward an everted upper lip, and relative prognathism. Patients may present with short stature, scoliosis, dental abnormalities, and brachydactyly with clinodactyly at 5th and digital syndactyly between the 3th and 4th toe. Ocular abnormalities may be present, such as deep-set eyes and close-spaced, synophrys, strabismus and iris abnormalities. The main otoryngological alterations are recurrent ear infections resulting in hearing loss, middle/inner ear abnormalities and deep hoarse voice.

Cognitive impairment and developmental delay are pronounced, however the most pronounced neurological alteration is sleep modifications. Patients with SMS often exchange nocturnal sleep for daytime naps, with changes of the circadian cycle and alterations in the release of melatonin [26-28].

Alterations of behavior are atypical and draw the most attention, because they are often unique to patients with SMS. Besides hyperactivity and attention seeking, patients with SMS may present agressive and self-injurious behavior, including hand biting, head banging, face slapping, self-hanging, onychotilomania and polyembolokotonia [16,19, 29-35]. Other signs reported in up to 50% of patients include obesity, cardiacs defects, seizures, cleft lip/palate and male hypogonadism [22].

http://dx.doi.org/10.5772/55721

Figures 1-7 are patients diagnosed with SMS in Genetic Counseling Service Dept Genetics. IBB/UNESP- Botucatu, Brazil, and several of these patients were published (Published with permission from *Genetics and Molecular Research-Online Journal*, Gamba et al., 2011. Genet. Mol. Res. 10 [4]: 2664-2670).

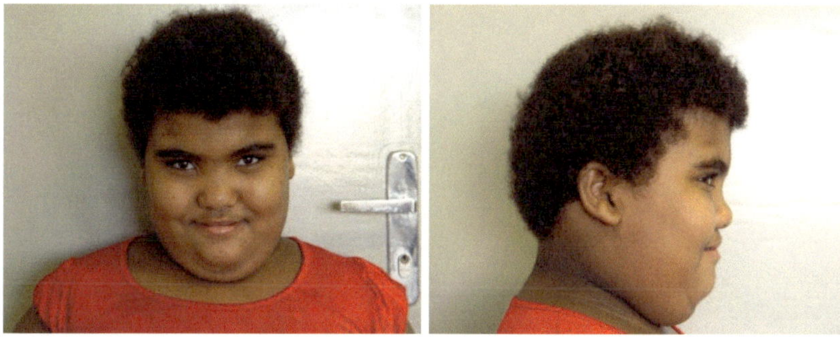

Figure 1. Patient 1. – 8.25 years-old

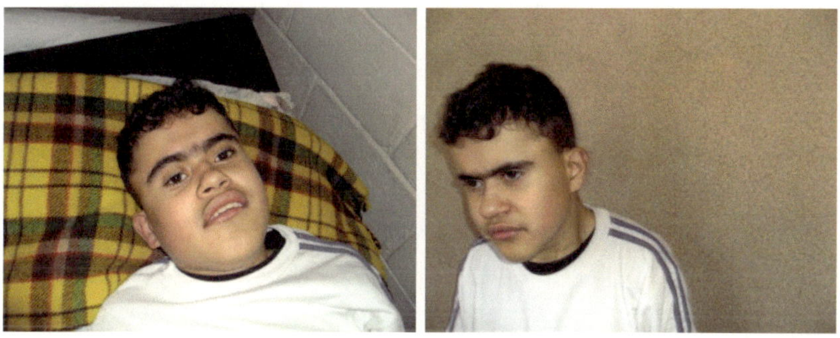

Figure 2. Patient 2 – 18.83 years-old

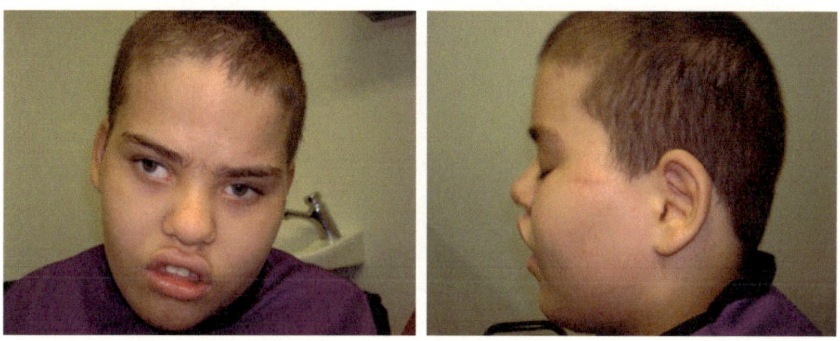

Figure 3. Patient 3. – 12.83 years-old

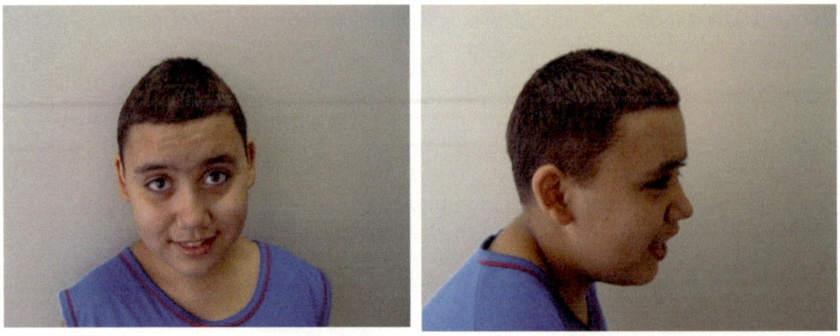

Figure 4. Patient 4 – 12.83 years-old

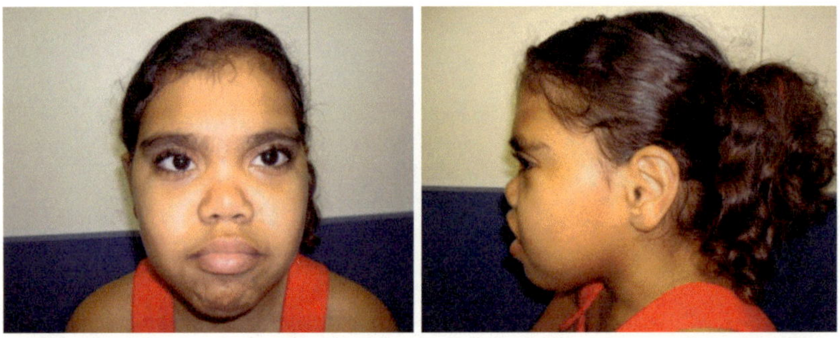

Figure 5. Patient 5. – 13.33 years-old

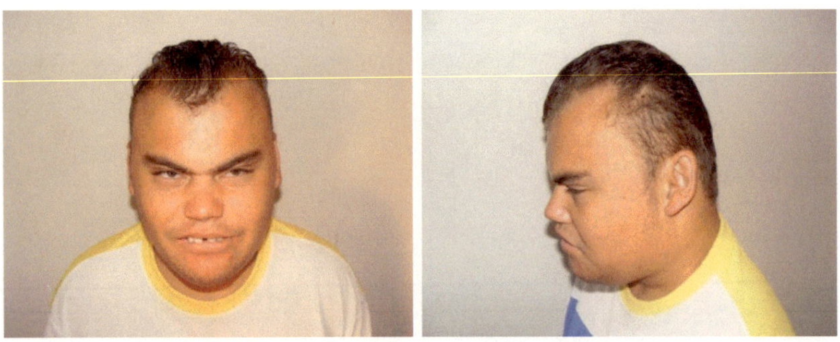

Figure 6. Patient 6 – 19.0 years-old

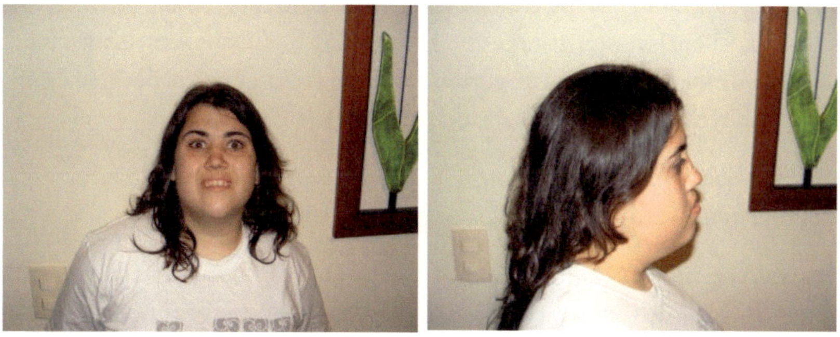

Figure 7. Patient 7. – 20.58 years-old

3.1.3. Adulthood

Adults with SMS have a diminution of stereotypic movements, but when frustrated, develop aggressive speech, with shouting or profanity at high volume. Little data is published regarding the life expectancy of patients with SMS. However it is believed that life expectancy is normal or similar to that of other individuals with the same level of cognitive dysfunction [36,37].

4. Genetics

Most cases of SMS are caused by a microdeletion on 17p11.2 that encompasses multiple genes, including the retinoic acid-induced 1, *RAI1*, gene. This deletion is observed in 90% of the cases, although in 10% of cases a point mutation in the *RAI1* gene is observed [2, 4, 6, 20, 38]. SMS microdeletions are caused by irregularities in chromosomal recombination mediated by repeat elements referred to as Low copy number repeats (LCR). Already [39] identified three copies of an LCR as being responsible for the deletion on 17p11.2. These repeats (LCRs proximal, middle, and distal - SMS REPs) form substrates for inter- and intrachromosomal recombination. In 70% of SMS cases, unequal meiotic crossovers result in nonallelic homologous recombination between the proximal and distal SMS REPs and a deletion of approximately 3.7Mb. In the remaining 30%, deletions are due to alternate SMS REPs (distal x medial). Moreover, AT-rich repeats and *Alu* elements may act as homologous recombination substrates, and nonhomologous mechanisms can generate deletions of atypical deletions sizes [40-42].

5. Diagnosis

SMS is suspected in individuals presenting distinctive facial features, a behavioral phenotype and sleep disturbance. Initial clinical suspicion of the disorder is confirmed by the presence of a microdeletion in the p11.2 region of chromosome 17 or a mutation in the *RAI1* gene. The unique SMS behavioral phenotype including sleep disturbance, a hoarse voice, characteristic hands and feet, excellent long-term memory, good ability and focus with computers, self-injury scars and typical facial features are important clues to the diagnosis. Because SMS will rarely be the only possible clinical diagnosis, exams are key to diagnosis.

SMS diagnosis is confirmed by detecting 17p11.2 deletion using classic cytogenetic analysis, molecular cytogenetic analysis, or molecular genetic methods [20].

Cytogenetic analysis by GTG banding at the 550 band level or higher can detect deletions of approximately 4Mb, which account for 70% of the cases. However fluorescent *in situ* hybridization (FISH) using an RAI1-specific probe is the most frequently used technique [20, 22, 43-44] (Fig.8 and Fig. 9).

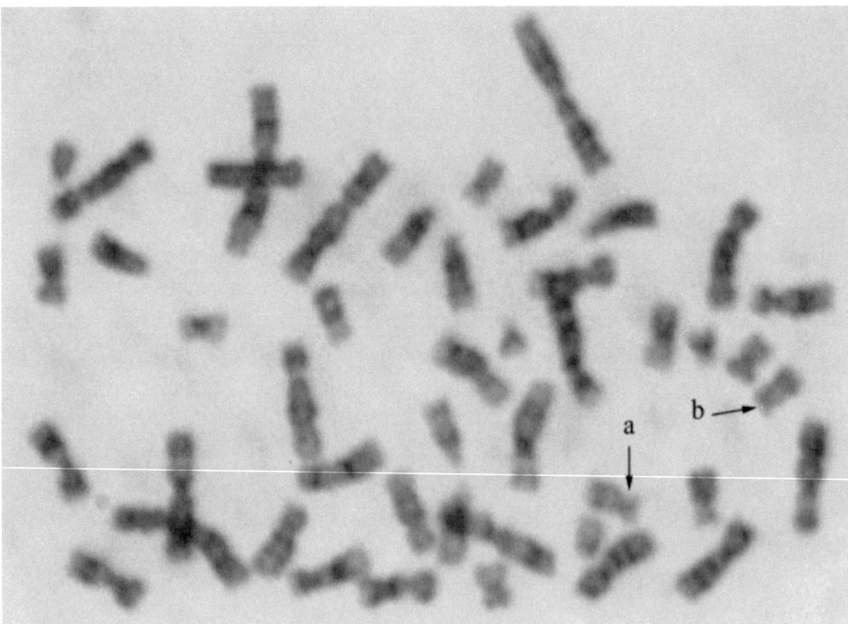

Figure 8. GTG banding of metaphase chromosomes showing the normal chromosome 17 (a) and deleted chromosome 17 (b). *Courtesy of the Cytogenetics Laboratory - SAG / IBB-UNESP-Botucatu, SP-BRAZIL*

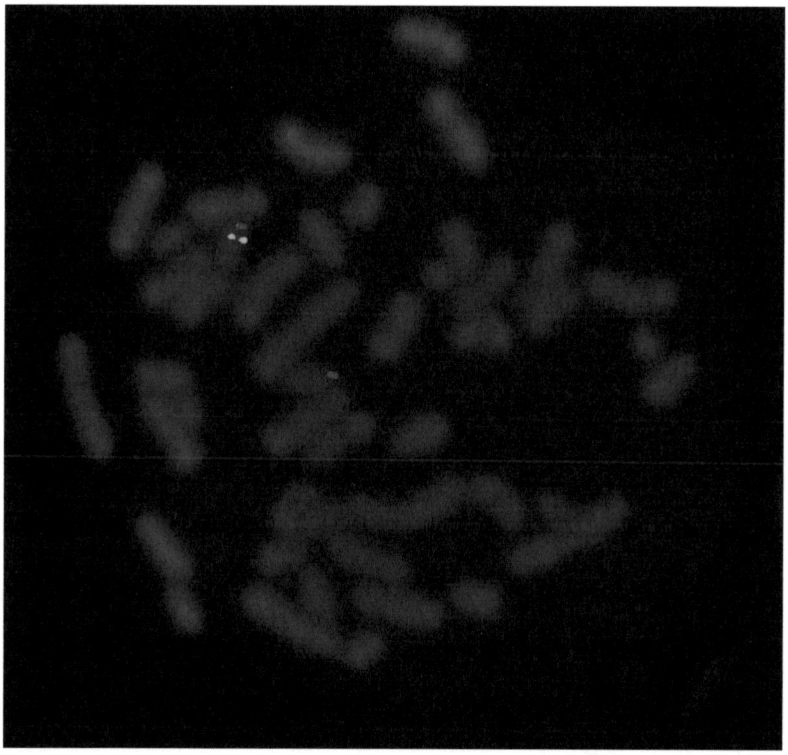

Figure 9. Metaphase FISH using the probe in the Smith-Majenis Cytocell ® (RAI 1 and Flii) / Miller-Dieker. The red area shows the control the green gene responsible for SMS. *Courtesy of the Cytogenetics Laboratory - SAG / IBB-UNESP-Botucatu, SP-BRAZIL*

Beyond these cytogenetic methods, methods that require only DNA for analysis are newer, cost-efficient, and can be used in a large number of patients at the same time. Additionally, MLPA or qPCR can identify smaller deletions at a higher resolution than FISH or G-banding [45].

6. Differential diagnosis

The differential diagnosis for SMS includes [13, 44].

- 1q36 deletion syndrome
- 9q34 deletion syndrome

As the clinical characteristics of SMS and these syndromes overlap, specific FISH tests are required for establishing a final diagnosis

- 22q11.2 deletion syndrome: velopharyngeal abnormalities and facial characteristics differentiate this syndrome from SMS

- Down syndrome: despite having several ovelapping features with SMS, Down syndrome can be diagnosed by simple karyotype analysis
- Williams-Beuren syndrome (WBS): SMS and this syndrome show opposing behavioral characteristics. While WBS patients are overfriendly, loquacious and frequently smiling, SMS individuals are shy, aggressive and restless
- Prader-Willi syndrome (PWS): although obesity may be present in both PWS and SMS, it is always of the morbid type in PWS patients:
- Sotos syndrome (SS): in SS patients, bone age is advanced while in SMS it is normal.

7. Treatment of manifestations

Patients with SMS present functional disturbances (obesity, sleep disturbances) and behavioral abnormalities (aggression, self-injury) which have prompted attempts to medically treat these alterations. Due to the low frequency of SMS, classical placebo-controlled prospective clinical drug trials have not been feasible. Clinical experience to date indicates that no drug is effective in alleviating any SMS symptoms in more than 60% of cases. There are a number of anecdotal reports of successful treatments, however many of unsuccessful treatments are likely unreported.

A review of pharmacological treatments with psychotropic drugs in patients with SMS was reported [46]. The medications were grouped into seven main categories: [1] stimulants; [2] antidepressants; [3] antipsychotics; [4] hypnotics; [5] mood stabilizers; [6] alpha 2 agonists; [7] and benzodiazepines. The stimulant category included methylphenidate, amphetamines, and others (e.g., pemoline). Antidepressants were subdivided into selective serotonin reuptake inhibitors (SSRIs), tricyclics (TCA), and others. The antipsychotic category was divided into typical and atypical. The hypnotic category included melatonin, diphenhydramine,and others. Mood stabilizers included ithium and anticonvulsants used for mood stabilization. Clonidine and guanfacine were grouped under alpha 2 agonists and all benzodiazepines were grouped together. The beta-blockers category was excluded due to a small number of reports. The study was conducted using medical histories of 62 patients with SMS. This study concluded that no consistent results were observed for any medicine or drug group, although the study did not exclude any of the drugs used.

Another elegant work [47,48] tested the effect of administration of B1-adrenergic antagonists together with melatonin in 10 patients with SMS, in an attempt to improve the circadian disturbances. These authors concluded that the administration of acebutolol in the morning and melatonin in the early evening allowed the biological clock reset and restore the normal rhythm of melatonin in SMS patients. The patients had improvements in sleep, diminution of naps during the day, with a higher state of attention and diminution of aggressive behavior.

8. Conclusions

Significant overlap between SMS's clinical features with other similar syndromes does makes it very difficult establish a clinical diagnosis. However, the uniqueness of the behavioral features of this condition should lead health care providers to request specific FISH testing. Treatment for SMS is merely relies on managing the symptoms. Individuals with SMS often require several forms of support, including physical therapy, occupational therapy, speech therapy, and particularly behavioral therapy, which are most effective if started early in life. Therefore, having an early diagnosis can help guide a person's health care through life, and open the doors to a network of information from professionals and other families dealing with the syndrome.

Author details

Danilo Moretti-Ferreira

São Paulo State University – Unesp, Bioscience Institute – Genetics Department, Botucatu, SP, Brazil

References

[1] Smith ACMMcGavran L, Waldstein G. Deletion of the 17 short arm in two patients with facial clefts.(1982). Am J Hum Genet 34 (Abstract- ASHG).

[2] Smith, A. C, Mcgavran, L, Robinson, J, Waldstein, G, Macfarlane, J, Zonona, J, Reiss, J, Lahr, M, Allen, L, & Magenis, E. Interstitial deletion of (17)(p11.2) in nine patients. Am J Med Genet. (1986). , 11.

[3] Potocki, L, Shaw, C. J, Stankiewicz, P, & Lupski, J. R. Variability in clinical phenotype despite common chromosomal deletion in Smith-Magenis syndrome. Genet Med. (2003). , 5, 430-4.

[4] Vlangos, C. N, Yim, D. K, & Elsea, S. H. Refinement of the Smith-Magenis syndrome critical region to approximately 950kb and assessment of 17deletions. Are all deletions created equally? Mol Genet Metab. (2003). , 11.

[5] Lupski JR: Genomic disorders: structural features of the genome can lead to DNA rearrangements and human disease traitsTrends Genet (1998).

[6] Slager, R. E, Newton, T. L, Vlangos, C. N, Finucane, B, & Elsea, S. H. Mutations in RAI1 associated with Smith-Magenis syndrome. Nat Genet. (2003). , 33, 466-468.

[7] Bi, W, Saifi, G. M, Shaw, C. J, Walz, K, Fonseca, P, Wilson, M, Potocki, L, & Lupski, J. R. Mutations of RAI1, a PHD-containing protein, in nondeletion patients with Smith-Magenis syndrome. Hum Genet. (2004). , 115, 515-24.

[8] Girirajan, S. Elsas Ii LJ, Devriendt KH, Elsea SH. RAI1 variations in Smith-Magenis syndrome patients without 17deletions. J Med Genet. (2005). , 11.

[9] Vilboux, T, Ciccone, C, Blancato, J. K, Cox, G. F, Deshpande, C, Introne, W. J, & Gahl, W. A. Smith ACM, Huizing M. Molecular Analysis of the Retinoic Acid Induced 1 Gene (RAI1) in Patients with Suspected Smith-Magenis Syndrome without the 17Deletion. PLoS ONE. (2011). e22861., 11.

[10] Vieira, G. H, Rodriguez, J. D, Carmona-mora, P, Cao, L, Gamba, B. F, & Carvalho, D. R. de Rezende Duarte A, Santos SR, de Souza DH, DuPont BR, Walz K, Moretti-Ferreira D, Srivastava AK. Detection of classical 17deletions, an atypical deletion and RAI1 alterations in patients with features suggestive of Smith-Magenis syndrome. Eur J Hum Genet. (2012). , 11.

[11] Greenberg, F, & Guzzetta, V. Montes de Oca-Luna R, Magenis RE, Smith AC, Richter SF, Kondo I, Dobyns WB, Patel PI, Lupski JR. Molecular analysis of the Smith-Magenis syndrome: a possible contiguous-gene syndrome associated with del(17)(Am J Hum Genet. (1991). , 11.

[12] Smith, A. C, & Duncan, W. C. Smith-Magenis syndrome: a developmental disorder of circadian dysfunction. In: Butler MG, Meaney FJ, eds. Genetics of Developmental Disabilities. Boca Raton, FL: Taylor and Francis Group; (2005). , 2005, 419-75.

[13] Smith ACMBoyd K, Elsea SH, Finucane BM, Haas-Givler B, Gropman A,Johnson KP, Lupski JR, Magenis E, Potocki L, Solomon B. (2010). Smith-Magenis Syndrome. updated 2010 Jan 7. In: Pagon RA, Bird TD, Dolan CR, Stephens K, editors. GeneReviews (Internet). Seattle (WA): University of Washington, Seattle., 1993-2001.

[14] Tomona, N. Smith ACM, Guadagnini JP, Hart TC. Craniofacial and dental phenotype of Smith-Magenis syndrome. Am J Med Genet. (2006). , 140, 2556-61.

[15] Udwin, O, Webber, C, Horn, y, & Abilities, I. and attainment in Smith-Magenis syndrome. Development Medicine and Child Neurology.(2001). , 43, 823-828.

[16] Dykens, E. M, & Smith, A. C. Distinctiveness and correlates of maladaptive behaviour in children and adolescents with Smith-Magenis syndrome. J Intellect Disabil Res. (1998). , 42, 481-489.

[17] Smith, A. C, Dykens, E, & Greenberg, F. Behavioral phenotype of Smith-Magenis syndrome (del 17Am J Med Genet. (1998). , 11.

[18] Sarimski, K. Communicative competence and behavioural phenotype in children with Smith-Magenis syndrome. Genet Couns. (2004). , 15, 347-355.

[19] Gropman, A. L, Duncan, W. C, & Smith, A. C. Neurologic and developmental features of the Smith-Magenis syndrome (del 17Pediatr Neurol. (2006)., 11.

[20] Elsea, S. H, & Girirajan, S. Smith-Magenis syndrome. Eur J Hum Genet. (2008)., 16, 412-421.

[21] Williams, S. R, Girirajan, S, Tegay, D, Nowak, N. J, Hatchwell, E, & Elsea, S. H. Array comparative genomic hybridization of 52 subjects with a Smith-Magenis-like phenotype: identification of dosage-sensitive loci also associated with schizophrenia, autism, and developmental delay. J Med Genet. (2009)., 47, 223-9.

[22] Gamba, B. F, Vieira, G. H, Souza, D. H, Monteiro, F. F, Lorenzini, J. J, Carvalho, D. R, & Moretti-ferreira, D. Smith-Magenis syndrome: clinical evaluation in seven Brazilian patients. (2011). Genet. Mol. Res., 10(4), 2664-2670.

[23] Einspieler C; Hirota H; Yuge M; Dejima S; Marschik PBEarly behavioural manifestation of Smith-Magenis syndrome (del 17in a 4-month-old boy. Developmental Neurorehabilitation. (2012)., 11.

[24] Wolters, P. L, Gropman, A. L, Martin, S. C, Smith, M. R, Hildenbrand, H. L, & Brewer, C. C. Smith ACM. Neurodevelopment of children under 3 years of age with Smith-Magenis syndrome. Pediatr Neurol (2009)., 41, 250-258.

[25] Hildenbrand HL; Smith ACAnalysis of the sensory profile in children with smith-magenis syndrome. Phys Occup Ther Pediatr. (2012)., 32(1), 48-65.

[26] Smith, A. C, Dykens, E, & Greenberg, F. Sleep disturbance in Smith-Magenis syndrome (del 17 Am J Med Genet. (1998)., 11.

[27] Boone, P. M, Reiter, R. J, Glaze, D. G, Tan, D-X, Lupski, J. R, & Potocki, L. Abnormal circadian rhythm of melatonin in Smith-Magenis syndrome patients with RAI1 point mutations. Am J Med Genet Part A. (2011)., 155, 2024-2027.

[28] Williams SR; Zies D; Mullegama SV; Grotewiel MS; Elsea SHSmith-Magenis Syndrome Results in Disruption of CLOCK Gene Transcription and Reveals an Integral Role for RAI1 in the Maintenance of Circadian Rhythmicity. Am J Hum Genet. (2012)., 90, 941-949.

[29] Finucane, B. M, Konar, D, Haas-givler, B, Kurtz, M. B, & Scott, C. I. The spasmodic upper-body squeeze: a characteristic behavior in Smith- Magenis syndrome. Dev Med Child Neurol. (1994)., 36, 78-83.

[30] Dykens, E. M, Finucane, B. M, & Gayley, C. Brief report: cognitive and behavioral profiles in persons with Smith-Magenis syndrome. J Autism Dev Disord 1997;Finucane, B.M. and Jaeger, E.R. Smith-Magenis syndrome. Ophthalmology, (1997)., 27, 203-211.

[31] Martin, S. C, Wolters, P. L, & Smith, A. C. M. Adaptive and maladaptive behavior in children with Smith-Magenis syndrome. Journal of Autism and Developmental Disorders, (2006). , 36, 541-552.

[32] Hicks, M, Ferguson, S, Bernier, F, & Lemay, J. F. A case report of monozygotic twins with Smith-Magenis syndrome. J Dev Behav Pediatr. (2008). , 29, 42-46.

[33] Smith, M. R, Hildenbrand, H, & Smith, A. C. Sensory motor and functional skills of dizygotic twins: one with Smith-Magenis syndrome and a twin control. Phys Occup Ther Pediatr. (2009). , 2009, 239-57.

[34] Heinze EG; Villaverde ML; López EM; Magro TC; Moura LF; Fernández M; Sampaio AFuncionamiento cognitivo general y habilidades psicolingüísticas en niños con síndrome de Smith-Magenis. Psicothema (2011). , 23, 725-731.

[35] Sloneem J; Oliver C; Udwin O; Woodcock KAPrevalence, phenomenology, aetiology and predictors of challenging behaviour in Smith-Magenis syndrome. J Intell Disab Research. (2011). , 55, 138-151.

[36] Osório A; Cruz R; Sampaio A; Garayzábal E; Carracedo A; Fernández-Prieto MCognitive functioning in children and adults with Smith-Magenis syndrome. Eur J Med Genet. (2012). , 55, 394-399.

[37] Elsea SH; Stephen RWilliams SR. Smith-Magenis syndrome: haploinsufficiency of RAI1 results in altered gene regulation in neurologicaland metabolic pathways. Expert Rev. Mol. Med. (2011).

[38] Chen, K. S, Manian, P, Koeuth, T, Potocki, L, Zhao, Q, Chinault, A. C, Lee, C. C, & Lupski, J. R. Homologous recombination of a flanking repeat gene cluster is a mechanism for a common contiguous gene deletion syndrome. Nat Genet. (1997). , 17, 154-63.

[39] Shaw, C. J, & Lupski, J. R. Implications of human genome architecture for rearrangement-based disorders: the genomic basis of disease. Hum Mol Genet. (2004). Suppl 1):RR64., 57.

[40] Shaw CJ & Lupski JRNon-recurrent 17deletions are generated by homologous and non-homologous mechanisms. Hum Genet. (2005). , 11.

[41] Tug, E, Cine, N, & Aydin, H. A Turkish patient with large 17deletion presenting with Smith Magenis syndrome. Genet Couns. (2011). , 11.

[42] Vlangos, C. N, Wilson, M, Blancato, J, Smith, A. C, & Elsea, S. H. Diagnostic FISH probes for del(17)(p11.2) associated with Smith-Magenis syndrome should contain the RAI1 gene. Am J Med Genet A. (2005). , 11.

[43] Vieira, G, Rodriguez, J. D, Boy, R, et al. Differential diagnosis of Smith-Magenis syndrome: 1deletion syndrome. Am J Med Genet A. (2012). , 36.

[44] Truong HT; Dudding T; Blanchard CL; Elsea SHFrameshift mutation hotspot identified in Smith-Magenis syndrome: case report and review of literature. BMC Medical Genetics (2010).

[45] Laje, G, Bernert, R, Morse, R, & Pao, M. Smith, ACM.2010. Pharmacological Treatment of Disruptive Behavior in Smith-Magenis Syndrome. Am J Med Genet Part C Semin Med Genet. (2010). C:, 463-468.

[46] De Leersnyder, H, Bresson, J. L, De Blois, M. C, Souberbielle, J. C, Mogenet, A, Delhotal-landes, B, Salefranque, F, & Munnich, A. Beta 1-adrenergic antagonists and melatonin reset the clock and restore sleep in a circadian disorder, Smith-Magenis syndrome. J Med Genet. (2003)., 40, 74-8.

[47] De Leersnyder, H. Inverted rhythm of melatonin secretion in Smith-Magenis syndrome: From symptoms to treatment. Trends Endocrinol Metab, (2006)., 17, 291-298.

Chapter 4

Humans Walking on All Four Extremities With Mental Retardation and Dysarthric or no Speech: A Dynamical Systems Perspective

Sibel Karaca, Meliha Tan and Üner Tan

Additional information is available at the end of the chapter

http://dx.doi.org/10.5772/55685

1. Introduction

1.1. Quadrupedal locomotion (walking on four extremities)

Locomotion is the movement of an organism from one place to another, often by the action of appendages such as flagella, limbs, or wings. In some animals, such as fish, a lumbering locomotion results from a wavelike series of muscle contractions (The American Heritage® Science Dictionary, 2005). Walking is travelling by foot; gait is the manner of locomotion; running is the act of travelling on foot at a fast pace; crawling is a slow mode of hand-knee or hand-foot locomotion. Walking on all four extremities (quadrupedal locomotion, QL) is the trait of the quadruped animals. Non-primate mammals usually utilize lateral-sequence QL, in which the hind limb touchdowns are followed by the ipsilateral forelimb touchdowns (symmetric gait). On the contrary, the non-human primates usually utilize a diagonal-sequence QL, in which the hind-limb moves with the contralateral forelimb in a diagonal couplet (asymmetric gait). Interestingly, only the animals exhibiting the diagonal-sequence QL with symmetrical gait evolved towards species with enlarged brains associated with highly complex neural circuits, till the emergence of human beings. The animals exhibiting lateral-sequence QL did not show such a phylogenetic progress compared to those with diagonal-sequence QL. Figure 1 shows the differences between lateral-(left) and diagonal-sequence (right) patterns of QL.

1.2. Evolutionarily preserved neural networks for QL

With regard to the origins of the diagonal-sequence QL, it is reasonable to conclude that the neural circuits for this kind of locomotion existed even in the most primitive tetrapods, lived

Figure 1. Difference between walking styles of primate (right) and non-primate (left) mammals. Most non-primate animals utilize lateral-sequence locomotion, but most primates utilize diagonal-sequence locomotion. Notice the filled *vs* unfilled extremities during lateral (left) and diagonal (right) locomotion and the interference between the fore- and hind limbs on the left side during diagonal gait.

in the Devonian period during transition from water to land. Thus, this type of locomotion is indeed phylogenetically the oldest locomotor trait, not the lateral-sequence QL. Namely, fossils due to the first fish-like tetrapods, lived approximately 395 MYA, were recently discovered on the Polish coast. From the fossil tracks left by a tetrapod, it was concluded that this most primitive quadruped animal walked with diagonal strides (see Figure 2), reflecting the lumbering locomotor movements as their ancestors lived in marine environments [1]. Interestingly, the quintessence of this diagonal-sequence coordination of the extremities during QL, did not change throughout the course of evolution, through salamanders and tuataras [2], to the emergence of non-human primates and even human beings during upright locomotion on two extremities [3, 4]. On the other hand, these results also suggest that the neural circuits responsible for the diagonal-sequence QL have been preserved for about 400 million years since the first emergence of QL in the fishlike tetrapods. In accord, the lumbering locomotor movements of tetrapods may even be visualized in human infants during crawling (see Figure 3), reasonably resulting from the activity of the ancestral neural networks still functioning approximately 400 million years later from the first emergence of the fishlike tetrapods. In accord, there are reports in the scientific literature supporting these considerations, i.e., the neural networks controlling the diagonal-sequence QL have been preserved throughout the evolutionary development for at least 395 million years since the emergence of the tetrapod-like fishes lived during the Devonian period [1,5,6].

As mentioned above, the neural networks playing a role in the emergence of the QL have been preserved for at least 395 million years since the Devonian tetrapod-like fishes [1]. Accordingly, it was reported an evolutionarily conserved daldh2 intronic enhancer in the frog, mouse, and chicken, being also involved in the formation of the neural tube throughout vertebrate species [7]. This evolutionary conservation of the enzyme playing a role in shaping the neural tube is essentially related to the evolutionary conservation of the neural networks for the diagonal-sequence QL. The mechanisms of this evolutionary preservation of the basic neural networks remain, however, unresolved. The genetic and/or epigenetic mechanisms contributing to the evolutionary development would shed some light on this subject.

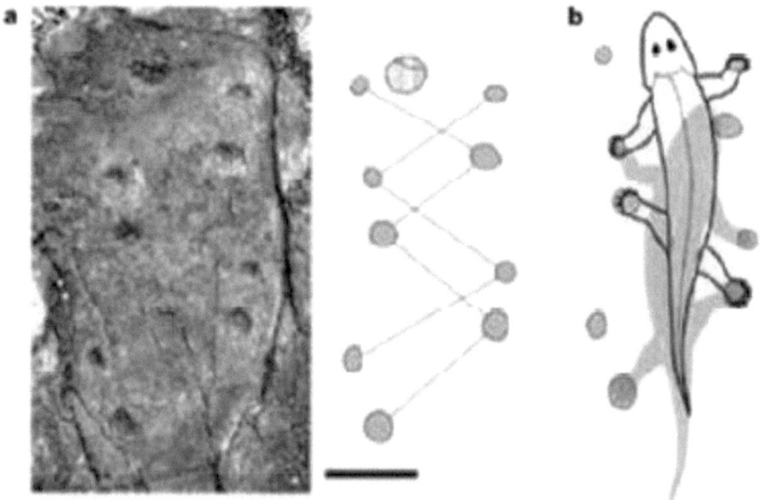

Figure 2. The trackway on the left (a) shows the hand and foot shapes in a diagonal stride pattern with a generic Devonian tetrapod fitted to this tackways (b). Notice the lumbering diagonal-sequence QL of this tetrapod, similar to its ancestral forms living in water.

2. Human beings with QL; Üner Tan Syndrome (UTS)

2.1. History

A human being habitually walking on all four extremities (quadrupedalism) was first discovered by Childs [8], the notable British traveler and writer, nearly a hundred years ago, on the famous Bagdad road near Havsa/Samsun on the middle Black Sea coast, at time of Ottoman Empire. This man probably belonged to a Greek family, since this region was populated by Greeks during this time. And, he probably was the son of a consanguineous family living in this closed Greek population with a high probability of consanguineous marriages. Childs described this man on page 29 of his book as follows:

"As we rose out of the next valley a donkey and a figure on the ground beside it attracted my attention...the figure moved in a curious fashion, and I went up to look more closely. And now it appeared I had fallen into the trap of a beggar...He sprang up and asked for alms, and because there were not immediately forthcoming went on all fours and showed a number of antics, imitating a dog and goat and other animals to admiration. Then I saw he was without thighs; that the knee-joint was at the hip, the leg rigid, and only half the usual length."

Fig. 4 shows the man walking on all four extremities reported in 1917 for the first time in the scientific literature by Childs, a famous British traveler and writer.

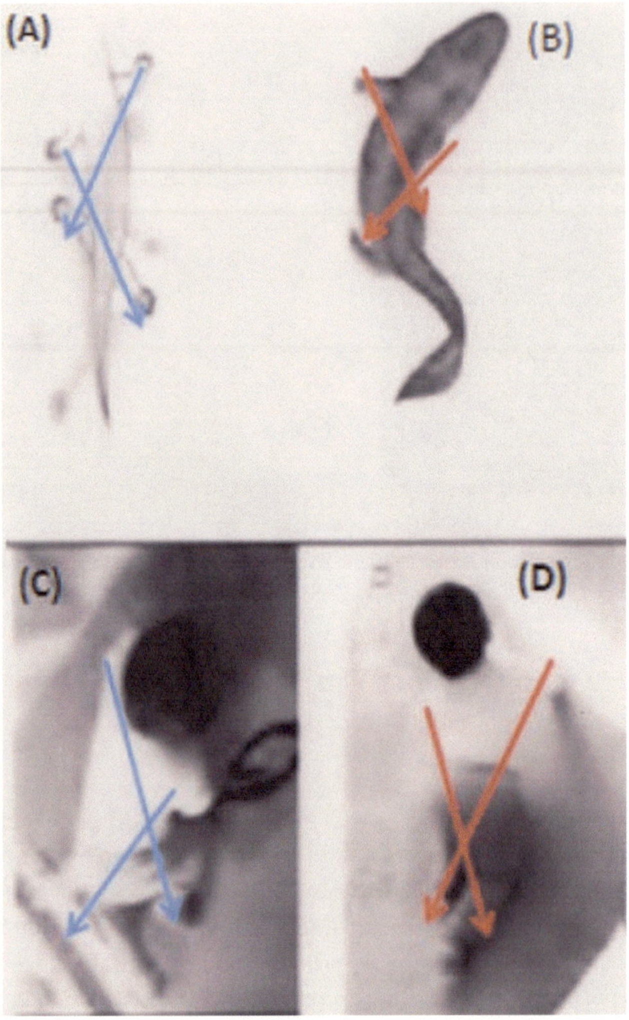

Figure 3. Diagonal-sequence QLs (see arrows) with lumbering in a generic Devonian tetrapod during the time of transition from water to land (A), proposed picture of a most primitive tetrapod, *Acanthostego,* with lumbering diagonal-sequence QL, a half fish, half reptile, lived in a swamp, 360 million years ago (MYA), see B. C, D: a contemporary human child crawling on all fours with accompanying lumbering movements, more or less similar to the very primitive Devonian animal lived almost 400 MYA.

Since the first discovery of a man with quadrupedalism, in 1917, not a similar human being was reported in the scientific literature, until the first description of five consanguineous kindred, resident in Southern Turkey, who exhibited a symptom complex with habitual QL, mental retardation, and dysarthric speech with limited conscious experience [9-15]. This pointed out a novel syndrome, which was referred to *Uner Tan Syndrome (UTS)* in some books, book chapters, and journal articles, accentuating the discovery's name. This novel syndrome was first reported, in 2005, in front of the members of the Turkish Academy of Sciences, in Ankara and Istanbul [9,10], also being published in some Turkish and English journals [11-15,

Figure 4. First man habitually walking on all four extremities, who was first discovered in Turkey during Ottoman Empire, in 1917, by Childs, the famous British traveler and writer, in Havsa/Samsun, Middle Black-Sea coast. The man was standing beneath his donkey, carrying torn trousers and shirt, pointing out his poverty.

see for reviews [16, 17]. This remarkable syndrome soon sparked a world-wide interest (see for reviews [18,19,], mainly because of its relation to human evolution.

2.2. UTS type-I and type-II

According to the childhood hypotonia in skeletal muscles, two subgroups of cases exhibiting the UTS could be distinguished: (i) UTS Type I, which included UTS cases without childhood hypotonia. Of 32 cases hitherto discovered in Turkey, 25 (78.1%) patients had UTS Type I. (ii) UTS Type II, which included UTS cases with childhood hypotonia. This was observed in 7 (21.9%) cases, who had no early ambulation during childhood. However, the childhood hypotonia disappeared and replaced with normal muscle tonus accompanied with QL during adolescence (see for reviews [16,17]. Figure 5 illustrates some of the UTS Type-I cases exhibiting diagonal-sequence QL, with coincidence of limbs and feet on the same side (interference effect), which would be disadvantageous for proper walking and running on all four extremities.

2.3. Forelimb and hind limb weight supports during QL

Non-primate mammals usually support their body weight more on the forelimbs than their hind limbs during QL [20,21]. Contrarily, most primates support their body weight more on their hind limbs than their forelimbs during QL [20,22,23]. The decreased forelimb weight support in non-human primates was interpreted as an adaptation to reduce stress on the forelimb joints, and facilitate the forelimb motility, especially for arboreal locomotion [24,25]. However, the human beings with QL without arboreal habits showed similar body mass distribution on the footfall patterns, i.e., less support on their hands (24% of their body weight) than their feet (76.0% of their body weight). Similarly, monkeys also support less than 30.0% of their body weight on their forelimbs [24,26,28]. According to Reynolds [24], 30-45% of the body weight was exerted on the forelimbs during QL of eight primates. Thus, the human beings with habitual QL also support their body weight more on their hind limb than their forelimbs, like their close relatives, the non-human primates. This body weight support pattern on hands and feet in human quadrupeds is consistent with the hypothesis that less body weight exerted on the hands than on the feet would be beneficial for fine manual skills in primates. A complete freeing of hands of human beings due to upright walking would be entirely associated with their highly developed hand skills and their accompanying co-development of their brains, all resulting from the replacement of a weight carrying function with a cognitive function of human hands. The severe mental retardation associated with the habitual QL in UTS cases could, therefore, be considered as an evolutionary example for the coupling between locomotor and mental abilities. The close coupling between the manual skill and cognitive ability was previously reported in humans [29-31] and great apes [32], consistent with Tan's psychomotor theory [33].

Figure 5. UTS cases walking on all four extremities exhibiting diagonal-sequence QL. Notice the interference between arms and legs on the same side due to diagonal sequence QL. Straight lines, forward and dotted lines, backward motions of the contralateral extremities.

2.4. Neurological examinations

All of the UTS-cases could understand simple questions and demands, but 24 cases (75.0%) in 7 families had no expressive speech at all, so that they replied the simple questions with one or two simple sounds. Only 8 cases (25.0%) had a dysarthric speech with a very limited vocabulary. Cognitive tests showed a severe mental retardation in all cases. Brain MRI scans revealed a cerebello-vermial hypoplasia with mild gyral simplification in cerebral cortex, except one case with normal cerebellum and impaired peripheral vestibular system instead of the central vestibular system in other cases. Truncal ataxia was present in all cases, but muscle tone was normal with strong arms and legs in these adult cases. The results of the neurological examinations, MRI and PET scans, and cognitive tests were presented in two review articles [16,17]. The clinical characteristics of the affected cases of the families from Turkey are summarized in Table 1.

2.5. Gender differences

There were 19 men and 13 women with UTS from Turkey, 2 men and 1 women from Morocco, 4 brothers from Brazil, 2 men from Iraq, 1 man from Mexico and 1 man from Chile. The number of men (n = 29, 67.4%) exceeded that of women (n = 14, 32.6%), the difference (34.8%) being, however, only marginally significant ($\chi^2 = 3.34$, df = 1, p =.07).

2.6. Cognitive tests

All of the patients exhibited severe mental retardation, according to the results of two cognitive tests. "Mini Mental State Examination Test" (MMSE), also known as the "Folstein test" [34], consisting of a 30-point questionnaire, testing for the individuals' attention, calculation, recall, language and motor skills, showed severe mental retardation in all of the cases (range = 0 to 2 points). The healthy siblings of the affected individuals were relatively normal in the MMSE test, with scores ranging between 25 and 29 points, although tey all shared the same environment. The Wais-R (Wechsler Adult Intelligence Scale-Revised) showed also severe mental retardation in the UTS cases, who obtained "0" to "4" points of a total 30 points. The results of the MMSE test are summarized in Table 2.

2.7. Genetics

The UTS is genetically heterogeneous. Namely, we found missense mutations in the following genes: VLDLR in Canakkale and Gaziantep families [35], WDR81 in Iskenderun family [36], ATP8A2 in Adana family [37]. Interestingly, the mother of the affected siblings in the Iskenderun family had type-I diabetes, which may be associated with developmental malformations, such as caudal regression in mice [38]. The VLDLR gene is involved in the controlling neuroblast migration in the developing central nervous system, see [35]. This gene shows an evolutionary conservation for at least 200 million years [39]. WDR81 gene is evolutionarily highly conserved trans-membrane protein, which is highly expressed especially in cerebellum and corpus callosum, see [36]. However, different mutations in a single gene may lead to different expressions of the same phenotype (allelic heterogeneity). Moreover, *similar genetic*

Findings	**Iskend. Type-I**	**Adana Type-I**	**Antep Type-I**	**Canak. Type-I**	**Kars Type-I**	**Afyon Type-I**	**Diyarb. Type-II**
N (QL)	6	3	7	4	2	3	7
Men	2	2	5	2	2	2	4
Women	4	1	2	2	0	1	3
Age	21-35	29-39	14-48	24-64	44-45	12-24	9-27
Mutation	17p.13	13q.12	9p.24	9p.24	(?)	9p.24	(?)
Ves.Imp.	Central	Pripheral	Central	Central	Central	Central	Central
Cerebel.	Hypopl.	Normal	Hypopl.	Hypopl.	Hypopl.	Hypopl.	Hypopl.
Vermis	Hypopl.	Normal	Hypopl.	Hypopl.	Hypopl.	Hypopl.	Hypopl.
Cer.cort.	Gy.simp	Normal	Gy.simp	Gy.simp	Gy.simp	Gy.simp	Gy.simp
DTR upp	Normal	Normal	Normal	Normal	Normal	Normal	Normal
Strength	Normal	Normal	Normal	Normal	Normal	Normal	Normal
Babinski	+ (3/6)	Absent	+ (3/7)	Absent	+ (1/2)	+ (1/3)	+ (1/7)
Tremor	Mild	Mild	+ (1/7)	(-)	(-)	(-)	(-)
Nystag.	(+)	(+)	(-)	(-)	(-)	(-)	+ (2/7)
E.Hypoto	(-)	(-)	(-)	(-)	(-)	(-)	(+)
Men.ret.	Severe	Severe	Severe	Mild	Severe	Severe	Severe
Speech	Dysarth.	Dysarth.	No	No	No	No	No
Standing	Yes	No (1/3)	Yes	Yes	Yes	Yes	Yes
Bip.walk	(+)	(-):(1/3)	(+)	(+)	(+)	(+)	(-):(1/7)

Table 1. Findings from families with UTS. Iskend.: Iskenderun (mutation in WDR81 gene), Adana (mutation in ATP8A2 gene), Canak.:Canakkale (mutation in VLDLR gene), Afyon (mutation in VLDLR gene), Diyarb.:Diyarbakir. Ves.Imp.:vestibular impairment; Cerebel.:cerebellum; Cer.cor.: cerebral cortex; DTR upp.: upper extremity deep tendon reflexes; DTR low: lower extremity deep tendon reflexes; M.tone: muscle tone; Nystag.:nystagmus; E.hypoto.: extremity hypotonia; Men.ret.: mental retardation; Bip.walk.: bipedal walking.

lesions can have entirely different phenotypes [40]. Thus, mutations in a single gene like VLDLR cannot be solely associated with quadrupedal locomotion in human beings. Given the genetic heterogeneity of the UTS, a specific gene directly responsible for the emergence of human quadrupedalism does not seem to be reasonable. Moreover, the missense mutations found in the affected cases may also be involved in neural functions other than the QL. Accordingly, missense mutation in VLDLR gene was also associated with congenital cerebellar hypoplasia [41], along with Norman-Roberts syndrome, characterized by microcephaly, hypertonia, hyperreflexia, severe mental retardation, and agyric cerebral cortex [42]. The VLDLR gene works with a protein, reelin, which is also associated with disorders such as Alzheimer's disease, schizophrenia, and bipolar disorder.

Questions	Patients' answers
What is today's date? What is the month? What is the year? What is the day of the week today? What season is it?	***Orientation in time:*** They gave unrelated answers such as 80,90,house,cow,dog, or did not give an answer at all, except thinking.
Whose house is this? What room is this? What city are we in? What country are we in?	***Orientation to place:*** Nobody could give a correct answer, or they replied with unrelated words such as summer, me, winter, cow, dog, mother, father, etc.
Repeat:ball,flag,tree	***Immediate recall:*** Only a few of them could recall these words.
Count backwards from 100 by 7	***Attention:*** They even could not count forwards from 0 to 10.
Recall 3 words I asked previously	***Delayed verbal recall:*** Nobody could recall the words previously asked
Name these items: watch, pencil	***Naming:*** Only some of them could name these items
Repeat following: No if, ands, or buts	***Repetition:*** Nobody could repeat them
Take the paper in your hand, fold it in half, and put it on the floor.	***3-stage command:*** Nobody could follow this command, and only some of them took the paper in the hand.

Table 2. Questions from the MMSE test and patients' answers.

Merlberg et al [43] re-evaluated the disequilibrium syndrome [44] in Swedish patients with non-progressive cerebellar ataxia, dysarthria, short stature, childhood hypotonia, and mental retardation, without VLDLR mutation. Interestingly, MRI showed a spectrum from normal to severe cerebellar hypoplasia. Similarly, the UTS cases of the Adana family did not exhibit cerebellar hypoplasia [13-15], despite a missense mutation in the ATP8A2 gene [37], suggesting no genetic association of the cerebellar hypoplasia in UTS.

Taking together, the genetic associations hitherto reported for the UTS seem to have no or only minor explanatory power, if any, for the origins of human quadrupedalism. In accord, Hall [45] argued how genetics failed to find solutions for the discrepancies concerning the so-called genetic diseases: *evidence is growing that your DNA sequence does not determine your entire genetic fate.. Larger scale genomic studies over the past five years or so have mainly failed to turn up common genes that play a major role in complex human maladies.* This argument seems to be also true for well-known disorders including diabetes, schizophrenia, and cancer, as Maher [46] stated:

even when dozens of genes have been linked to a trait, both the individual and cumulative effects are surprisingly small and nowhere near enough to explain earlier estimates of heritability.

3. Darwinian medicine; UTS

Why are only some rare cases predisposed to walk on all four extremities similar to our early ancestors? The quest to find an answer to this and similar questions was the starting point for establishing a new discipline, *"Darwinian medicine"*, which is a novel concept providing a foundation for all medicine [47]. The aim of the Darwinian medicine is the evolutionary understanding of aspects of the body with regard to its vulnerability to disease(s), as Zampieri [47] stated: *It tries to find evolutionary explanations for shared characteristics that leave all people vulnerable to a disease.* Evolutionary or Darwinian medicine may be useful to understand better why diseases exist despite natural selection [48,49]. A number of diseases were considered as Darwinian disorders, such as tuberculosis, Huntington's disease, depression, obesity, anxiety, pain, nausea, cough, fever, vomiting, fatigue, epilepsy, obsessive compulsive disorder, and schizophrenia [50-53]. In this framework, the UTS with the reappearance of the ancestral features such as quadrupedal locomotion and primitive cognition including no speech in most of the cases, and severe mental retardation may also be considered as a further example related to Darwinian medicine.

Rapoport [54] first introduced the concept of *"phylogenic diseases"*, such as Alzheimer's disease as a "phylogenic regression", comparing brain aging involution to the reversed phenomenon of Darwinian evolution [55. Accordingly, many neurodegenerative diseases such as Parkinson's disease, schizophrenia, Alzheimer's disease, and many highest level gait disorders including the UTS with ancestral QL, i.e., the re-emergence of old automatism of pre-human gait, may also be considered under these phylogenic diseases. In this context, the recently introduced paleoneurologic standpoint may help us to more deeply understand the pathogenesis of the neuropsychiatric diseases, provided that they are reconsidered under evolutionary perspective.

4. Complex systems; Self-organization

The word "complex" may be defined as "consisting of interconnected or interwoven parts" [56]. Many complex systems have the tendency to spontaneously generate novel and organized forms, such as ice crystals, galactic spirals, cloud formation, lightning flashes in the sky, or polygonal impressions in the earth. The spontaneously generated formation in the nature are in no way designed by anything, not even by natural selection, being entirely the art of nature with self-organizing properties within complex systems, following the principle, *the sum of the parts is greater than the parts taken*

http://dx.doi.org/10.5772/55685

independently, Contrary to Isaac Newton's arguments, *...the whole is the sum of all the parts..*

Complex systems have a strong tendency to self-organize, i.e., spontaneous formation of patterns in open nonequilibrium systems. This is also the quintessence of all living systems.

For instance, insects can spontaneously build their nests or hives, hunt in groups, and explore the food resources in their environment. The evolution of the biological forms and structures may also be associated with self-organization. Some authors have even questioned the centrality of natural selection in evolution, since Darwinism essentially ignores the principles of self-organization. Accordingly, Oudeyer [57] argued: *Thus, the explanation of the origins of forms and structures in the living can not only rely on the principle of natural selection, which should be complemented by the understanding of physical mechanisms of rom generation in which self-organization plays a central role.*

Self-organization is closely coupled with "emergence", a fundamental property of complex systems, which is the unpredictable product of the system, resulting from interconnections and interactions between parts of a dynamical system; entities, interactions, and the environment are key contributors to emergence [58]. The main characteristic of the UTS, human quadrupedalism, may also be an emergent property of the locomotor development. In accord with the dynamical systems theory, and the principles of self-organization, it can be stated that no genetic or neural code may be the causative factor for the emergence of the human quadrupedalism. As mentioned above, we could not isolate a single gene responsible only for the QL in human beings, minimizing the role, if any, of any genetic code in the emergence of human QL.

Human quadrupedalism may be considered as a strange attractor, a state of a dynamical system toward which that system tends to evolve. For instance, the EEG may exhibit one type of strange attractor while a person is at rest, but another type of strange attractor during mathematical thinking. The common property of strange attractors is their unpredictability. The rarely occurring locomotor pattern in the UTS cases, i.e., QL, may be related to its unpredictability as a strange attractor. An entirely different locomotor strange attractor emerged in a man from Tanzania, who exhibited all of the symptoms of the UTS, including truncal ataxia and no upright ambulation with mental retardation and no speech. However, an entirely novel and unpredictable locomotor pattern emerged in this man as a strange attractor. Namely, his QL was upside down, i.e., in face-up position. He used his hands and feet for QL, but used palms and heels instead of the soles (see Fig. 6). This is the first reported case exhibiting the UTS with inverse QL.

In essence, the dynamical systems tend to control the outcome of the system to find which patterns can possibly be built from the systems components to begin with, and the structural constraints of the environment, the self-organizing phenomena being basic mechanisms for the emergence of any adaptive behavior, such as the adaptive self-organization phenomena playing a role in the developmental emergence of the human quadrupedalism.

Figure 6. Tanzanian man with UTS, walking on all four extremities but with inverse quadrupedal locomotion: further example for a locomotor strange attractor.

5. Central Pattern Generators (CPG)

The locomotor system is closely related to CPGs embedded in the spinal cord, which is a set of motoneurons responsible for locomotion [59], probably also involved in the human QL. The spinal motor system seems to be similar in all quadrupeds and human beings [4]. Individuals with or without UTS may all share the same neural networks responsible for the diagonal-sequence quadrupedal locomotion as the nonhuman primates, because they all are using the common neuronal control mechanisms for locomotion [60]. However, the CPGs do not reflect the coordinated walking pattern in intact animals, since there are separate CPGs for each leg in the cat [61]. On the other hand, the presence of the CPGs in higher primates is much less convincing. This could be due to the increased role of the corticospinal tractus in primates, suppressing the spinal motor circuitry responsible for relatively rough locomotor movements, and facilitating the skilled hand movements. According to Duysens et al [61], CPGs have no direct equivalent in human beings.

The concept of CPGs did not find supporters among system theoreticians. For instance, Thelen and Smith [62] argued the notion of the CPGs as the essence of locomotion does not fit the data…They simply do not account for what we really observe in developing organisms..The fact of development is not explained by a list of innate ideas. Just as the assumption of a built-in CPG does not explain the development of walking. These authors further stated that real data from real frogs, chicks, cats, and humans render the construct of the CPG illusory.. If the program contains the instructions for the entire sequence of behaviors ahead of time, how can novel and adapted forms be generated? Actually, the CPGs exhibit one of the principles of biological self-organization as dynamic entities, i.e., different neural networks may induce similar outcomes, while similar neural networks can produce different outcomes. Namely, the CPGs are not static, previously hard-wired, firmly organized neural networks, they are rather loosely organized systems under the influence of the steadily changing chemical or sensory control, with newly emerged functional circuits [63]. Moreover, Neuronal networks within CPGs can change itself according to current conditions and exhibit transitions between functional states, resulting from dynamic instabilities occurring within the system with dynamic interactions at the neuronal, synaptic, and network levels [63].

6. Maturation theories

The concept of motor development showed a gradual shift from the traditional neuronal maturation theory towards the dynamic systems theory. It was believed, in the mid-1990s, that the development of the central nervous system occurred through the genetically predetermined neural networks and spinal reflexes, under the control of the cerebral cortex. Accordingly, the locomotor actions such as standing and walking in infants would result from the gradual maturation of the CNS under the influence of the cerebral cortex, not learnt by experience. The traditional maturation theory utilized the longitudinal studies to show the developmental sequence of motor behaviors in infants and young children. This was mainly elaborated by Gessel [64], Shirley [65], and McGraw [66,67], who searched for rules governing the order of changes during motor maturation. Konner [68] stated that *motor development sequences are largely genetically programmed.* The development of early motor behaviors was attributed to the maturation of the cortico-spinal pathways [69,70]. Despite some valuable information gained from the traditional maturation theory, it was far from explaining the dynamics of locomotor development. In this context, Ulrich [71] argued that *it is not at all clear how genetic codes can be translated into even simple patterned neural organization...behavior is much more than a simple neural pattern* (p.321).

Contrary to the traditional maturation theories, the contemporary approaches considered the properties of complex systems with many dynamically interacting subcomponents, to be able to solve the problems related to locomotor development. This dynamic systems theory considers the behavior of a system, not by taking it as separate parts, but by taking these parts to see under which circumstances they dynamically cooperate to produce the whole behavioral pattern such as locomotor functions. According to the dynamic systems theory, the behavioral patterns can emerge from the dynamic interactions of multiple subsystems; genetic or neural codes are not represented *a priori* in the brain, nor are locomotor patterns, such as walking and running. The emergence of locomotion is a self-organizing process, as in other complex systems. According to Ulrich [71, p.324], *the coordination pattern emerges spontaneously and is self-organized and opportunistic.* Taking together, there are two major but current and conflicting theories involved in the development of locomotor control: neuronal maturationist theory and the dynamic systems theory. According to the first theory, the maturation of the CNS occurred through the genetically predetermined neural networks; the locomotor development results from progressively maturated and hence increased cortical control on the spinal reflexes. Controversially, however, the system theoreticians did not accept the neural-maturationist theories, asking how can the timetable of motor solutions be encoded in the brain or in the genes. Accordingly, Kelso et al [72] utilized the dynamic systems theory to better explain the developmental emergence of locomotion in human beings. These authors argued that a behavior, such as a locomotor pattern may result from the combined dynamic actions of, for instance, muscle strength, body weight, postural support, motivation, and brain development, in addition to the environmental initial conditions and task requirements.

7. Neuronal group selection theory

In addition to the neuronal maturation theory and the dynamic systems theory, there is a third theory, the neuronal group selection theory (NGST) [73], which combines the "nature" part of the neural-maturationist theories with the "nurture" part of the dynamic systems theory. The neuronal groups are collections of many neurons interconnected by excitatory and/or inhibitory synapses as well as recurrent feedback circuits. According to the NGST, the structural and functional characteristics of these neuronal groups are determined by evolution. During locomotor development, behavior and experience produce afferent information for the central nervous system, which is used for the neuronal selection, according to the strength of the synaptic connections. The changed connectivity allows for a situation-specific selection of neuronal groups, which can be adapted to environmental constraints. The NGST emphasizes the role of the complex information processing originating from an intertwining of information from genes and the environment. This is not consistent with the "nature-nurture" debate. During motor development in early fetal life, the spontaneous fetal movements (primary variability), i.e., the self-generated motor activity with the consequent self-generated afferent information, may explore all the locomotor possibilities within the neurobiological and anthropometric constraints within the CNS, preserved during evolution.

During postnatal development, all of the intentional motor behaviors are within the frame of "primary variability". The neuronal networks emerging during this developmental phase, especially prominent in the cerebral cortex are suitable for the selection of the appropriate locomotor circuits, responsible, for instance, for the infantile crawling. The most effective motor pattern gradually emerges following exploratory continuous information processing within the CNS. The time-sequence for the selection process changes with function, for instance, the second half year after birth for arm reaching. The postural activity of neck and truck muscles are direction specific before infants cat sir independently at about five months after birth. The most efficient selection for the postural balance occurs around 12-18 months of age.

The long duration of the developmental processes suggests that long-lasting motor experiences are needed for the establishment of the secondary neuronal networks. This may be associated with the late-onset quadrupedalism in some UTS cases [74]. Actually, exercise may be beneficial for the selection of the most effective neuronal pattern, by reducing the amount of secondary variation [75]. The reverse occurs in the absence of exercise, similar to the UTS cases without exercise at all.

The NGST for the locomotor development is closely related to the concept of the adaptive self-organization. Namely, the developmental selection is the differential survival of developmental units, which was proposed as an explanation for examples of self-organization [76].

8. Dynamics of locomotor development in humans

The contemporary views on the ontogenic development of locomotor skills accentuate the role of the self-organizing processes within the scope of complex systems. As mentioned above, the neural patterns playing a role in the emergence of the diagonal-sequence QL have existed since about 400 MYA during the Devonian period, having arisen with the first appearance of the ancestral tetrapods. That is, this type of locomotion is indeed phylogenetically the oldest locomotor trait of tetrapods. Interestingly, the quintessence of this locomotor activity did not change during evolution through salamanders and tuataras [2], till the emergence of non-human primates and even human beings exhibiting diagonal-sequence movements between arms and legs, even during their upright walking [3]. It may thus be concluded that the neural generators responsible for the diagonal-sequence QL may already be present in the complex locomotor systems of primates, including humans. Taking together, it may be argued that the neural patterns responsible for the QL in human beings may emerge through exploration of available solutions within the CNS, such as the ancestral neural generators for the QL and then selection of preferred patterns, such as the CPGs [77-79]. Following this ontogenetic theory, it may be concluded that the emergence of the diagonal-sequence QL in human beings may be the result of a prenatal exploration and subsequent neuronal group selection process following the principles of the self-organizing dynamic systems [80].

The cases exhibiting UTS seem to be unable to make the secondary selection for the neural networks appropriate for bipedal locomotion during infantile development. That is, they could not make the transition from the infantile stage of crawling on all fours to upright standing and bipedal walking. Their brain apparently explored the possible solutions for locomotion, but could not select the neural patterns for bipedal locomotion, because of the structural anomalies in their brain. Instead, their brain could select only one ancestral locomotor pattern available for their locomotion, which was already present since about 400 MYA. This is the ancestral neural network responsible for the diagonal-sequence QL, emerged during the Devonian period of evolution. This gait unstable initially apparently becomes stable with practice during childhood, so that they later move with great ease, speed, and well-developed balance. On the other hand, the locomotor self-organizing process may take a long time in some UTS cases with late emergence of QL at about puberty, see [16,17], a period associated with hormonal changes with beneficial effects on the motor system, accelerating the self-organizing processes, resulting in the emergence of a most suitable locomotor pattern to travel around, walking on all four extremities.

9. UTS vs socio-economic status

Neither the complex systems including the self-organizing processes nor the neuronal group selection mechanisms alone can be realized without considering the dramatic influence of the environmental factors on the holistic processes occurring in the emergence of the UTS. Namely, the single environmental factor shared by all of the cases was their extremely poor living conditions due to their very low socio-economic status. Accordingly, all of the hitherto discovered UTS cases all over the world lived in poverty, resident in developing countries. According to the Databank of the World Bank, the mean GDP (gross domestic product) of the developed countries where no UTS cases were found was 4520.00 US$, whereas the mean GDP of the developing countries where all of the UTS cases were found was 1202.7 US$. The above results suggest that the UTS is a disease of poverty. In other words, UTS with human quadrupedalism and severe mental retardation may be triggered by the environmental factor, low socio-economic status. In this context, the rates of the developmental disorders are almost twice as high in the poorer countries and in the lower income populations than the higher income groups. Over 80% of cases with intellectual disabilities are living in low- and middle-income developing countries [81] Actually there is strong relationship between poverty and common mental disorders, which were found to be about twice as frequent among the poor people compared to rich people [82], where some cases with UTS except Turkey were also found to be resident.

The malnutrition, due to low-income socio-economic status, may cause epigenetic changes, leading to impaired prenatal development of the CNS. A close association between epigenetic status as measured by global DNA methylation and socio-economic status was indeed recently reported [83]; the global DNA hypomethylation was associated with the most deprived group of individuals, compared to the least deprived group [84]. The close relationship between the epigenetic status and the socio-economic status may also be applied to the UTS, a multifactorial-complex disorder, similar to other neuropsychiatric and neurodegenerative diseases. Epigenetics refer to modifications in gene activity without changing the original DNA sequence, depending upon the environmental clues. Similar to other neurodegenerative diseases, the UTS may also comprise multifactorial processes, such as genetic, epigenetic, and environmental components [85] There are consistent reports suggesting the epigenetic mechanisms are responsive to environmental exposures during both pre- and post-natal development in humans. With regard to the most effective environmental factor, the under-nutrition due to low socio-economic status, Heijmans et al [86] reported that persons prenatally exposed to famine showed epigenetic changes compared to their unexposed same-sex siblings. Apparently, these results are consistent with the hypothesis that the triggering factor for the emergence of the UTS with quadrupedalism, mental retardation, and impaired speech may be the under-nutrition, which may detrimentally affect the pre- and postnatal psychomotor development through changing the epigenetic mechanisms.

http://dx.doi.org/10.5772/55685

10. Concluding remarks

The first man habitually walking on all four extremities was discovered nearly a hundred years ago in Turkey by the famous British traveler and writer, Childs, on the Black Sea coast, near Samsun, on the famous Baghdad Road, during the time of Ottoman Empire. After a silent period lasting for almost 100 years, in 2005, 6 cases with habitual quadrupedal locomotion (QL) were described in Southern Turkey. These individuals exhibited a never-before-described syndrome with habitual quadrupedal locomotion, severe mental retardation, and dysarthric speech without conscious experience, mostly cerebello-vermian hypoplasia and mildly simplified cortical gyri, referred to *Uner Tan Syndrome (UTS).* The number of men exceeded the number of women at p =.07 level, suggesting a male preponderance in the UTS. The syndrome showed genetic heterogeneity.

UTS can be considered within the framework of the autosomal recessive cerebellar ataxias, associated with different genetic mutations, such as the disequilibrium syndrome, Cayman ataxia, and Joubert syndrome. These closely related syndromes show overlapping symptoms, such as truncal ataxia, psychomotor delay, and dysarthric speech. These syndromes also show genetic heterogeneity, which is shared by many diseases. Thus, genetics alone cannot be informative for the origins of many syndromes, including the UTS. This is consistent with the dynamical systems theory, with the essential argument there may not be a single element, such as a genetic and/or a neural code, that predetermines the emergence of human quadrupedalism. Rather, the self-organizing processes occurring within a complex system may be involved in the developmental origins of the UTS, consisting of many decentralized and local interactions among neuronal, genetic, epigenetic, and environmental subsystems.

UTS was considered in two subgroups: Type-I and Type-II, the former exhibiting persistent early-onset QL without infantile hypotonia, the latter exhibiting late-onset QL with early-onset hypotonia in skeletal muscles. Comparison with other closely related syndromes such as dysequilibrium syndrome, Cayman ataxia, and Joubert syndrome, suggested that UTS may be differentiated from other similar ataxic syndromes by exhibiting early- or late-onset QL, no hypotonia in skeletal muscles, and no short stature, contrary to severe hypotonia without ambulation, and short stature, among others, in related syndromes, see [16].

Similar to non-human primates, but contrary to non-primate species, the UTS cases utilized the diagonal-sequence quadrupedal locomotion to travel around. The evolutionary advantage of this type of locomotion is obscure. Interestingly, however, only primates with this evolutionarily primary locomotor trait followed an evolutionary route favoring the emergence of higher primates till the human beings. The non-primate mammals with lateral-sequence QL did not follow such a phylogenetic progress. The diagonal-sequence QL was phylogenetically oldest type of locomotion, since the first tetrapods within the Devonian period utilized this kind of locomotion. This suggests that the neural networks for the diagonal-sequence QL were reserved during the evolution from first tetrapods till human beings since about 400 MYA.

A remarkable advantage of the primates with diagonal-sequence QL was that only they could utilize their hands for fine manipulations, freed from weight-bearing functions following erect posture and bipedal walking. The reduced body weight support on hands than feet in non-human primates and human beings with habitual QL (see above) would be beneficial for the development of fine uni- and bi-manual motor skills.

It was suggested that UTS may be considered as a phylogenetic regression in light of Darwinian medicine, associated with an evolutionary understanding of disorders using the principles of evolution, such as natural selection. In some UTS cases, prominent supraorbital tori were observed in cranial MRIs, more or less similar to those in non-human primates. In addition to the diagonal-sequence QL and the body weight support predominantly on the hind limbs more than the forelimbs, this was taken consistent with the theory of evolution in reverse, i.e., the reappearance of a lost function or structure that was typical of remote ancestors.

The developmental emergence of the human QL was related to the self-organizing processes occurring in complex systems, selecting one preferred behavioral state or locomotor trait out of many possible attractors. Since the dynamic systems provide enormous flexibilities in this respect, this is an unpredictable event. With regard to locomotor patterns, the dynamical systems of the developing child may prefer or create some kind of locomotion, resulting from interactions of the internal components and the environmental conditions, without a direct role of any causative factors, such as genetic and/or neural codes. The developmental emergence of human locomotion including QL is a developmental event in which the self-organization processes play the major role, no innate or previously prescribed codes being essential for the emergence of walking during locomotor development. In UTS with impaired balance, the system will find the most suitable and most comfortable, and hence preferred, mode of locomotion, spontaneously generating novel and organized forms and attractor states. These spontaneously occurring unpredictable attractors may result in the emergence of the face-down or face-up diagonal-sequence QL. In light of the dynamical systems theory, the contribution of single factors such as genetic and/or neural codes to the emergence of these locomotor patterns were rejected, considering the current scientific research in these fields, which are consistent with the concept of self-organization, suggesting no single element has causal priority.

The low socio-economic status leading to malnutrition in all UTS cases, all of them being from developing countries, was suggested as a triggering factor for the epigenetic changes occurring during the pre- and post-natal development of the brain. Namely, under-nutrition may trigger epigenetic changes in the brain, affecting the primary variability, in the first phase of locomotor development. In fetuses undergoing to epigenetic changes, the developing brain is then influenced by the aberrant proprioceptive information from fetus, resulting in impaired outcome of the developing brain, associated with psychomotor retardation and selection of the evolutionarily preserved neuronal groups with ancestral locomotor networks, leading a so-called reverse evolution in bipedal locomotion.

With regard to the neuronal group selection theory, the neural system can explore all motor possibilities by means of the self-generated, spontaneous motor activity, and with consequently self-generated afferent information transmission to CNS. The selection of the neuronal groups within the ancestral neural networks in the CNS, available since about 400 MYA, may then lead to the motor development in the next phase, i.e., the neuronal group selection by experience during infancy. In UTS cases, this phase of the locomotor development would stop because of the unavailability of the neuronal groups contributing to the postnatal emergence of bipedal locomotion, continuing ancestral locomotion on all four extremities, resulting from selection of the available ancestral neural networks for QL. So, the infants with UTS cannot select the appropriate neural networks for bipedal locomotion, since some of the neural structures necessary for the well-balanced upright locomotion are damaged in these infants, due to the cerebellar hypoplasia and cortical gyral simplification.

Following the phase of the ancestral neuronal groups responsible for human locomotion, the adaptive variability phase occurs at two to three years of age, with maturation in adolescence through experience. In cases with UTS within the same age range, this adaptive variability phase for bipedal locomotion cannot be accomplished, instead they keep the more primitive motor repertoires from the first variability and neuronal selection phase, resulting in persistence of the selection of the ancestral neuronal groups responsible for the very primitive diagonal-sequence quadrupedal locomotion, evolutionarily conserved since about 400 MYA.

Acknowledgements

This work was partly supported by the Turkish Academy of Sciences (Ankara, Turkey).

Author details

Sibel Karaca[1], Meliha Tan[1] and Üner Tan[2]

1 Başkent University, Adana Research and Training Center, Department of Neurology, Adana, Turkey

2 Çukurova University, Medical School, Department of Physiology, Adana, Turkey

References

[1] Niedzwiedzki, G, Szrek, P, et al. (2010). Tetrapod trackways from the early Middle Devonian period of Poland. *Nature*, , 463, 43-48.

[2] Reilly, S. M, Mcelroy, E. J, et al. (2006). Tuataras and salamanders show that walking and running are ancient features of tetrapod locomotion. Proc R Soc B, , 273, 1563-1568.

[3] Donker, S. F, Beek, P. J, et al. (2001). Coordination between arm and leg movements during locomotion. *J Mot Behav*, , 33, 86-102.

[4] Dietz, V. (2002). Human bipeds use quadrupedal coordination? *Trends Neurosci,* , 25, 462-467.

[5] Daeschler, E. B, Shubin, N. H, et al. (2006). A devonian tetrapod-like fish and the Evolution of the tetrapod body plan. *Nature,* , 440, 757-763.

[6] Shubin, N. H, Daeschler, E. B, & Jenkins, F. A. (2006). The pectoral fin of *Tiktaalik* and the origin of the tetrapod limb. *Nature,* , 440, 764-771.

[7] Castillo, H. A, Cravo, R. M, et al. (2009). Insights into the organization of dorsal spinal cord pathways from an evolutionarily conserved raldh2 intronic enhancer. *Development,* , 137, 507-518.

[8] Childs, W. J. (1917). *Across Asia Minor on Foot.* William Blackwood and Sons, Edinburgh and London.

[9] Tan, U. (2005a). Yeni bir sendrom (dört-bilekli yürüyüş, ilkel konuşma, mental retardasyon) ve insan Evrimi. *Turkish Academy of Sciences,* Ankara.

[10] Tan, U. (2005b). Üner Tan sendromu: insan ruhunun evrimine ilişkin yeni bir teori. *Turkish Academy of Sciences,* Istanbul.

[11] Tan, U. (2005c). Unertan sendromu ve insan ruhunun evrimine iliskin yeni bir teori. *Biyobank,* (3)

[12] Tan, U. (2005d). Unertan syndrome; quadrupedality, primitive language, and severe Mental retardation; a new theory on the evolution of human mind. *NeuroQuantology,* , 4, 250-255.

[13] Tan, U. (2006a). A new syndrome with quadrupedal gait, primitive speech, and severe mental Retardation as a live model for human evolution. *Int J Neurosci,* , 116, 361-369.

[14] Tan, U. (2006b). Evidence for "Unertan Syndrome" and the evolution of the human mind. *Int J Neurosci,* , 16, 763-774.

[15] Tan, U. (2006c). Evidence for "Unertan Syndrome" as a human model for reverse evolution. *Int J Neurosci,* , 116, 1539-1547.

[16] Tan, U. (2010). Uner Tan syndrome : history, clinical evaluations, genetics, and the dynamics of human quadrupedalism. *The Open Neurology Journal,* , 4, 78-89.

[17] Tan, U, Tamam, Y, Karaca, S, & Tan, M. (2012). Üner Tan syndrome: review and emergence of human quadrupedalism in self-organization, attractors and evolutionary perspectives. In: *Latest Findings in Intellectual and Developmental Disabilities Research,* Prof. Üner Tan (Ed.), 978-9-53307-865-6InTech Pub., Croatia.

[18] Downey, G. Quadruped: Uner Tan syndrome, part 1. Retrieved from http://blogs.plos.org/neuroanthropology/2010/09/03/human-quadruped-unertan syndrome-part-1/.

[19] Downey, G. (2010b). legs good, 4 legs better: Uner Tan syndrome, part 2, Retrieved from http://blogs.plos.org/neuroanthropology/2010/09/05/legs-good-4-legs-betteruner tan-syndrome-part-2/.

[20] Demes, B, Larson, S. G, et al. (1994). The kinetics of primate quadrupedalism: "hindlimb drive" reconsidered. *J Hum Evol,* , 26, 353-374.

[21] Schmitt, D, & Lemelin, P. (2002). Origins of primate locomotion : gait mechanics of the woolly opossum. *Am J Phys Anthropol,* , 118, 231-238.

[22] Kimura, T, Okada, M, & Ishida, H. (1979). Kinesiological chracteristics of primate walking: its significance in human walking. In: Morbek ME, Preuschoft H, & Gomberg N, editors. *Environment, behavior, and morphology: dynamic interactions in primates.* New York: G. Fischer. , 297-311.

[23] Schmitt, D. (2003). Insights into the evolution of human bipedalism from experimental studies of humans and other primates. *J Exp Biol,* , 206, 1437-1448.

[24] Reynolds, T. R. (1985). Stresses on the limbs of quadrupedal primates. *Am J Phys Anthropol,* , 67, 101-116.

[25] Larson, S. G. (1998). Unique aspects of quadrupedal locomotion in nonhuman primates. In: Strasser E, et al., editors. *Primate locomotion: recent advances.* New York: Plenum Press. , 157-174.

[26] Kimura, T, Okada, M, & Ishida, H. (1979). Kinesiological chracteristics of primate walking: its significance in human walking. In: Morbek ME, Preuschoft H, & Gomberg N, editors. *Environment, behavior, and morphology: dynamic interactions in primates.* New York: G. Fischer. , 297-311.

[27] Kimura, T. (1985). Bipedal and quadrupedal walking of primates: comparative dynamics. In Kondo S, editor. *Primate morpho-physiology, locomotor analyses and Human bipedalism.* Tokyo: University of Tokyo Press. , 81-105.

[28] Schmitt, M. D. (1994). Forelimb mechanics as a function of substrate type during quadrupedalism in two anthropoid primates. *J Hum Evol,* , 26, 441-457.

[29] Tan, U. (1989). Manual proficiency in Cattle's intelligence test in left-handed male and female subjects. *Int J Neurosci,* , 44, 17-26.

[30] Tan, U. (1990). Relation of spatial reasoning ability to hand performance in male and Female Left-handers to familial sinistrality and writing hand. *Int J Neurosci,* , 53, 143-155.

[31] Tan, U, Akgun, A, & Telatar, M. (1993). Relationships among nonverbal intelligence, hand speed, and serum testosterone level in left-handed male subjects. *Int J Neurosci,* , 71, 21-28.

[32] Byrne, R. W. (2003). The manual skills and cognition that lie behind hominid tool use. In *The evolution of thought; evolutionary origins of great ape intelligence.* (A.E. Russon

& D.R. Begun, Eds.). Cambridge University Press. Chapter DOI:http://dx.doi.org/10.1017/CBO9780511542299.005

[33] Tan, U. (2007). The psychomotor theory of human mind. *Int J Neurosci*, , 117, 1109-1148.

[34] Folstein, M. F, Folstein, S. E, & Mchugh, P. R. (1975). Mini-mental state". A practical method for grading the cognitive state of patients for the clinician. *J Psychiat Res*, , 12, 189-198.

[35] Ozcelik, T, Akarsu, N, Uz, E, et al. (2008). Mutations in the very low density Lipoprotein receptor VLDLR cause cerebellar hypoplasia and quadrupedal locomotion inhumans. *Proc Natl Acad Sci USA*, , 105, 4232-4236.

[36] Gulsuner, S, Tekinay, A. B, et al. (2011). Homozygosity mapping and targeted genomic Sequencing reveal the gene responsible for cerebellar hypoplasia and quadrupedal locomotion in consanguineous kindred. *Genome Res*, , 21, 1995-2003.

[37] Onat, O. E, Gulsuner, S, Bilguvar, K, Basak, A. N, Toplaoglu, H, Tan, M, Tan, U, Gunel, M, & Ozcelik, T. (2012). Missense mutation in the ATPase, Aminophospholipid transporter protein ATP8A2 is associated with cerebellar atrophy and quadrupedal locomotion. *Eur J Hum Gen*, doi:ejhg.2012.170.

[38] Chan, B. W, et al. (2002). Maternal diabetes increases the risk of caudal regression caused by retinoic acid. *Diabetes*, , 51, 2811-2816.

[39] Bar, I. Lambert de Rouvroit, C., Goffnet, A.M. ((2000). The evolution of cortical development. An hypothesis based on the role of the reelin signaling pathway. *Trends Neurosci*, , 23, 633-638.

[40] Prasun, P, Pradhan, M, & Agarwal, S. (2007). One gene, many phenotypes. *J Postgrad Med*, , 53, 257-261.

[41] Moheb, L. A, Tzschach, A, Garshasbi, M, et al. (2008). Identification of a nonsense mutation in the very low-density lipoprotein receptor gene (VLDLR) in an Iranian family with dysequilibrium syndrome. *Eur J Hum Genet*, , 16, 270-273.

[42] Sakai, J, Hoshino, A, Takahashi, S, et al. (1994). Sructure, chromosome location, and expression of the human very low density receptore gene. *J Biol Chem*, , 269, 2173-2182.

[43] Melberg, A, Örlén, H, et al. (2011). Re-evaluation of the dysequilibrium syndrome.*Acta Neurol Scand*, , 123, 2-33.

[44] Sanders, G. (1973). The dysequilibrium syndrome. A genetic study. *Neuropadiatrie*, , 4, 403-413.

[45] Hall, S. S. (2011). The genomes dark matter. *Tech Rev*, January/February, , 53-57.

[46] Maher, B. (2008). The case of the mising behavior. *Nature*, , 456, 18-21.

[47] Zampieri, F. (2009). Origins and history of Darwinian medicine. *Humana.Mente,* , 9, 13-38.

[48] Williams, G. W, & Nesse, R. M. (1991). The dawn of Darwinian medicine. *Q Rev Biol,* , 66, 1-22.

[49] Stearns, S. ed. ((1998). *Evolution in health and disease.* Oxford University Press. Stockinger, W., Brandes, C., et al. (2000). The reelin receptor ApoER2 recruits JNK-interacting proteins-1 and-2. *J Biol Chem,* , 275, 25625-25632.

[50] Abed, R. T, & Pauw, K. W. (1998). An evolutionary hypothesis for obsessive compulsive disorder : a-psychological immune system. *Behav Neurol,* , 11, 245-250.

[51] Eskenazi, B. R, Wilson-rich, N. S, & Starks, P. T. (2007). A Darwinian approach to Huntington's disease: subtle health benefits of a neurological disorder. *Med Hypothesis,* , 69, 1183-1189.

[52] Scorza, F. A, & Cysneiros, R. M. et. al. ((2009). From Galapagos to the labs: Darwinian medicine and epilepsy today. *Epilelpsy Behav,* , 16, 388-390.

[53] Pearlson, G. D, & Folley, B. S. (2008). Schizophrenia, psychiatric genetics, and Darwinian psychiatry: an evolutionary framework. *Schizophr Bull,* , 34, 722-733.

[54] Rapoport, S. I. (1988). Brain evolution and Alzheimer's disease. *Rev Neurol,*, 144, 7990.

[55] Ghika, J. (2008). Paleoneurology: neurodegenerative diseases are age-related diseases of specific brain regions recently developed by homo sapiens. *Med Hypotheses,* , 71, 788-801.

[56] Bar-yam, Y. (1997). *Dynamics of complex systems. Perseus* Books, Reading Massachusetts, USA.

[57] Oudeyer, P-Y. (2006). Self-organization: complex dynamical systems in the evolution of speech. In: *Self-organization in the evolution of speech,* P-Y. Oudeyer (ed.). Oxford studies in the evolution of language 6. Oxforf University Press.

[58] Dobrescu, R, & Purcarea, V. L. (2011). Emergence, self-organization and morphogenesis in biological systems. *J Med Life,* , 4, 82-90.

[59] Grillner, S, & Wallen, P. (1985). Central pattern generators for locomotion, with special reference to vertebrates. *Ann Rev Neurosci,* , 8, 233-261.

[60] Shapiro, L. J, & Jungers, W. L. (1994). Electromyography of back muscles during Quadrupedal and bipedal walking in primates. *Am J Phys Anthropol,* , 93, 491-504.

[61] Duysens, J. Van de Crommert, W.A.A. ((1998). Neural control of locomotion; Part 1: The central pattern generator from cats and humans. *Gait and Posture,* , 7, 131-141.

[62] Thelen, E, & Smith, L. B. (1994). *A dynamic systems approach to the development of cognition and action.* MIT press: Cambridge, Massachusetts, London, England.

[63] Kelso, J. A. S. (1995). Dynamic patterns. A Bradford Book, The MIT Press, England.

[64] Gessel, A. (1928). *Infancy and human growth.* New York: Macmillan.

[65] Shirley, M. M. (1931). The first two years: a study of 25 babies. *Postural and locomotor development.* Minneapolis: University of Minnesota Press., 1

[66] Mcgraw, M. B. (1932). From reflex to muscular control in the assumption of an erect Posture and ambulation in the human infant. *Child Dev,* , 3, 291-197.

[67] Mcgraw, M. B. (1943). *The neuromuscular maturation of the human infant.* (Reprinted 1990 as Classics in Developmental Medicine London: Mac Keith Press).(4)

[68] Konner, M. (1991). Universals behavioral development in relation to brain myelination. In: Gibson, K.R. & Petersen, A.C. (Eds.), *Brain maturation and cognitive development:comparative and cross-cultural perspectives.* New York: Aldine de Gruyter, , 181-223.

[69] Von Hofsten, C. (1984). Developmental changes in the organization of prereaching movements. *Dev Psychol,* , 20, 378-388.

[70] Jeannerod, M. (1988). *The neural and behavioral organization of goal-directed movements.* Oxford: Clarendon Press.

[71] Ulrich, B. D. (1997). Dynamic systems theory and skill development in infants and children. In: *Neurophysiology & Neuropsychology of motor development*. Connolly, K.J., & Forssberg, H. (Eds.), Mac Keith Press: London, , 321.

[72] Kelso, J. A, Holt, K. G, et al. (1981). Patterns of human interlimb coordination emerge from the properties of non-linear, limit cycle oscillatory processes: theory and data. *J Mot Behav,* , 13, 226-261.

[73] Hadders-algra, M. (2000). The neuronal group selection theory: promising principles for understanding and treating developmental motor disorders. *Dev Med Child Neurol,* , 42, 707-715.

[74] Tan, M, Karaca, S, & Tan, U. (2010). A new case of Uner Tan syndrome-with late childhood quadrupedalism. *Mov Dis,* , 25, 652-653.

[75] Pedotti, A, Crenna, P, et al. (1989). Postural synergies in axial movements: short and longterm adaptation. *Exp Brain Res,* , 74, 3-10.

[76] Frank, S. A. (1997). Developmental selection and self-organization. *Biosystems,* , 40, 237-243.

[77] Gibson, E. J. (1988). Exploratory behavior in the development of perceiving, acting, and the acquiring of knowledge. *Ann Rev Psychol,* , 39, 1-41.

[78] Sporns, O, & Edelman, G. M. (1993). Solving Bernstein's problem: a proposal for the development of coordinated movement by selection. *Child Dev,* , 64, 960-981.

[79] Thelen, E, & Corbetta, D. (1994). Exploration and selection in the early acquisition of skill. *Int Rev Neurobiol, ,* 37, 75-102.

[80] Chang, C-L, Kubo, M, et al. (2006). Early changes in muscle activation patterns of Toddlers during walking. *Inf Beh Dev, ,* 29, 175-188.

[81] Holzer, C. E, et al. (1998). The increased risk for specific psychiatric disorders among persons of low socio-economic status. *Am J Soc Psychiat, ,* 4, 259-271.

[82] Patel, V, et al. (1999). Women, poverty and common mental disorders in four Re-structuring societies. *Soc Sci Med, ,* 49, 1461-1471.

[83] Borghol, N, Suderman, M, et al. (2011). Associations with early-life socio-economic position in adult DNA methylation. *Int J Epid, ,* 41, 62-74.

[84] Mcguinness, D, Mcglynn, L. M, et al. (2012). Socio-economic status is associated with epigenetic differences in the pSoBid cohort. *Int J Epidemiol,* doi:10.1093/ije/dyr215.

[85] Marques, S. C. F, Oliveira, C. R, et al. (2011). Epigenetics in neurodegeneration: a new Layer of complexity. *Prog Neuropsychopharm Biol Psychiat, ,* 35, 348-355.

[86] Heijmans, B. T, Tobi, E. W, et al. (2008). Persistent epigenetic differences associated with prenatal exposure to famine in humans. *PNAS, ,* 105, 17046-17049.

Chapter 5

Sexuality and Sex Education in Individuals with Intellectual Disability in Social Care Homes

Stanislava Listiak Mandzakova

Additional information is available at the end of the chapter

http://dx.doi.org/10.5772/55782

1. Introduction

Sexuality is a natural part of person's life, which accompanies a person throughout his/her live. Unfortunately, rules concerning sexuality in individuals with intellectual disability (ID) are not the same as those imposed on the rest of society. Generally, sexuality in people with ID is considered a problem. There are still persistent myths and prejudice in this area related to incorrect thoughts, that individuals with ID are asexual or they remain eternal children or, on the contrary, they are sexually impulsive. Individuals with ID are often not allowed to express themselves in the area that has a sexual content.

This chapter aims to analyze specific aspects of quality-of-life in individuals with ID with regard to sexuality, partnerships, and sex education implementation in social care homes (SCHs), and to identify some of the determining factors. The current study does not deal only with "low-risk" sexual activities, e.g. kissing and cuddling but also with sexual activities that present pregnancy risk, disease transmission or injury of the sexual partners.

The purpose of the current study is a deeper insight and understanding of sexuality in individuals with severe intellectual disability (SID)[1] in SCHs, and sex education implementation in these people and aims to improve their life in SCHs.

2. The current status of knowledge — Theoretical and research findings

In Slovakia, the attention to the issue of sexuality in individuals with ID has been studied sporadically. Sexuality and partnerships in people with ID are very complex. Only few acquire such extent of independence that enables them to live with a natural partner and sex life without their parents or care assistant's supervision. Prevendárová [17] found that more than 80% of individuals with ID stay at the mental age of eight. According to this author [17], it is necessary to understand their sexuality manifestations, e.g. kissing or simply a feeling of closeness, in a similar way. A similar opinion has been expressed by Škorpíková [24] who believes that many individuals with SID do not desire or need an intercourse in their life. They find themselves in so-called pregenitality, which is a form of child sexuality. According to this author, people with SID prefer caressing, cuddling, and exhibition to an intercourse. Šurabová [24] has identified the following specific characteristics of sexuality in individuals with SID:

- Individuals with SID do not incline towards so-called total sexual behavior that involves coital contact. As for the erotic behaviors, tactile contacts and snuggling– rather than contact satiation, seems to be sufficient for them.
- If behaviors resembling masturbation occur, it is only a contact with one's genitals accompanied by pleasant feelings, but it is not a specifically conscious behavior.

From these observations, it can be concluded that, in comparison to the general public, individuals with SID generally do not possess offensive or active sexual behavior. However, not everyone agrees with these conclusions. For instance, Mellan [23] emphasizes that it is incorrect to assume that a need for sexual life is not created in individuals with ID. Sexuality could only be imperfectly expressed or delayed. Accordingly, Šedá [24] concludes that, an individual with SID cannot usually cope with his/her own sexuality, does not have an opportunity to satisfy it, and often does not know how to do it. In regard to the above-mentioned contradictory opinions, we find it necessary to study sexuality in individuals with SID. In the following sections we introduce some of the findings, in particular international research studies of sexuality and its manifestation in individuals with SID that can help us better understand this issue in terms of Slovak research.

There is a paucity of studies around the world evaluated in a large and representative sample concerning the question that what individuals with ID expect from sexuality, in terms of intercourse, emotions, marriage, and children upbringing. Our aim is to review and evaluate research studies carried out on the subject of sexuality in people with ID.

Perhaps because of society's attitudes to sexuality in individuals with SID, this topic has not been studied extensively. Studies of sexuality problems, and therefore of reproduction, have for many centuries been restricted in the European territory mostly by an influence of religious opinions. Perhaps it is not necessary to specify in detail, rigid, non-scientific, and still spread and fiercely defended opinions of the Church on questions of sex life, reproduction, and in particular attempts to control and regulate it with birth control methods. Despite the significant changes in people's attitude particularly after World War II, sexologists working in this field have identified areas around the world that still have no clear understanding of this issue, particularly if the results should threaten the existing attitudes. In the following section, we will conduct a review of some of sexuality research studies in individuals with SID that has formed the basis of our research.

Our investigation shows that many individuals in institutional care have experienced some sexual activities, even though the range of their sexual behavior is probably limited partially due to their sexual repression [23]. Timmers and colleagues [23] found that 65% of men and 82% of women with ID had had sex, although the frequency was much lower in comparison to others. Unfortunately, problems such as low sample size, inability to describe the seriousness of disability, and relying on intermediated reports about behavior of individuals with ID (not obtained by direct observation or statements of individuals with ID) weaken the strength of research studies in this field.

Conod and Servais [1] believe that expectations of individuals with ID regarding sexuality and its manifestation differ significantly in regard to the level of their ID. According to these authors, studies specifically concerning sexual activity in individuals with ID found that, 50% of women with moderate to severe ID, at age 11 to 23, had had sex. McGillivary and colleagues confirmed these findings [13] showing that, out of 60 interviewed adults (35 men and 25 women) with mild to moderate ID, 42% were sexually active. Similarly, Diederich and Graecen [3] via email interviews with institutions found that 41% of adults with SID in institutional care in the Paris area had already had sex. Using a similar methodology [5], it has been reported that 48% of directors of public residential institutions with less than 50% representation by individuals with SID point to the fact that sexual relations have occurred, whereas in 15%, it has been considered to be common. Obviously, these activities depend not only on expectations of individuals with ID, but also on opportunities provided by their social environment.

In a study by McGillivary [13], 82% of women with moderate ID living in coeducational institutions already had sex, whereas in non-coeducational institutions only 4% of individuals have had sex. Another finding points to the fact that 14.5% of individuals with mild and moderate ID without previous sexual experience claim to consider becoming sexually active as soon as an opportunity arises.

Pueschel and Scola [18] interviewed parents of 73 teenagers with Down syndrome (36 men and 37 women) and concluded that more than half of the teenagers expressed an interest in the opposite sex, masturbation was registered in 40% of men and 22% of women, only four parents stated that their son or daughter had ever mentioned desire for an intercourse. Contradictions with McGillivary's results are likely caused by methodological differences between the two studies (interviews with individuals with ID versus parents), as well as by individualities of people with Down syndrome in terms of their sexual expectations.

The knowledge of individuals with ID about sexuality and related topics differ in those living in institutional and non-institutional environments, but overall is very low [12]. The findings from the two studies in author's opinion suggest that:

- Young individuals living in coeducational environment have a slightly higher level of knowledge than those living in non-coeducational institutions. The more accurate predictor of cognition is mental age rather than IQ or real age.
- The young individuals with or without a handicap have a rather low knowledge level. Individuals with a handicap, however, have a bigger chance to obtain information from their contemporaries (often ill informed) than from reliable sources, such as books by experts. The access to the exact information is limited and it increases a need for complex programs of sex education for this population.

In many developing countries, the premarital sexual activity is much less socially acceptable and therefore it is likely to expect and assume a significant difference in sexual expectations also in people with ID.

Another possible factor influencing human sexual behavior is the gender. It has been shown that sexual behavior is influenced by biological factors as well as by the influence of parental upbringing, attitude of society to men and women and, by economical factors [18].

Next, we present an analysis of attitudes to sexuality in individuals with ID, as attitudes are very important elements and they influence our behavior and actions. In the context of our study, we refer to SCHs' employees, who are in touch with individuals with SID.

McCabe [12] studied attitudes of both parents and care assistants. The findings about care assistants' attitudes are contradictory – those working in institutions are often aware of the fact that sexual activity exists, although not every time they are able to acknowledge and deal with it. The author explains the differences in care assistants' responses by such factors as environment in which a care assistant works, the choice of questions asked, and the time and the country in which data are collected.

As for the Slovak research studies, we can refer to a survey by Števková [10] focused on attitudes to sexuality in individuals with ID. The results of this though non-representative survey suggests that emotional manifestations of individuals with ID are not accepted and there is absence of awareness.

At the end of this theoretical analysis, it can be concluded that the Kinsey Reports on human sexual habits based on personal interviews, formed an inestimable basis of sociological information about types of sexual manifestation. The evaluation of the work-study could considerably expand the research on sexual manifestations and sex education implementation in the general public and also in individuals with SID.

3. Research of the quality-of-life of individuals in SCHs in regard to sexuality, partnerships, and sex education

3.1. Research problems and objectives

The main Aim of our study was to analyze and describe particular aspects of quality-of-life in individuals with SID in the area of sexuality and partnerships, and identify factors determining the individuals' sexuality and sex education in SCHs.

Our main aim has been divided into four sub aims:

1. To identify quality indicators of social services provided by institutions in general (a size of an institution, coeducation, a number of individuals sharing a room, employees – a number, education, etc.).
2. To obtain new information about the sexuality in individuals with SID in terms of intensity and frequency of their occurrence.

3. To identify attitudes of employees to sexuality and sex education in individuals with SID and to describe differences in terms of their sexual education.

4. To study the level of awareness of SCHs' employees in sexuality and sex education in individuals with SID.

3.2. Research hypotheses and justification

In relation to research on individuals with SID behavior in the area of sexual content, we propose three hypotheses:

According to our hypotheses, specific aspects of sexuality will vary in terms of:

1. Gender; men with SID will show a significantly higher frequency and intensity of the sexuality in comparison to women with SID.

2. The severity of ID; individuals with moderate ID will show significantly higher frequency and intensity of the sexuality in comparison to individuals with SID.

3. Institution's coeducation; individuals of coeducational institutions will show a significantly higher frequency and intensity of the sexuality in comparison to individuals of non-coeducational institutions.

In respect of SCHs employees' attitudes to sexuality in individuals with SID, we have assumed that attitudes of employees to sexuality in individuals with SID will vary in terms of:

4. The presence of a training program in sex education; employees who have completed a training program in sex education will show significantly more positive attitudes to sexuality in individuals with SID in comparison to employees who have not completed a training program in sex education.

5. The attitudes of male employees; male employees will be significantly more positive to sexuality in individuals with SID than female employees.

6. The attitudes of employees to sexuality in general public; the more positive attitudes to sexuality in general, the more positive attitudes they will have to sexuality in individuals with SID.

7. The type of sexual behaviors; employees will show significantly more positive attitudes to non-reproductive sexual behavior than parenthood of individuals with SID.

3.3. Justifications

1. As Raboch [19] and Oakleyová [16] reported, the sexual behavior is influenced by biological factors, parental upbringing, general public attitudes to men and women, and economical factors. According to Weiss and Zvěřina [25], men mention higher need and frequency for sexual satisfaction than women. On the other hand, as the authors note,

there is only little evidence that differences in a sexual activity between men and women would be conditioned biologically.

2. Formulating hypothesis 2, there are studies that suggest sexual behavior in a person with ID correlates with the severity of the ID. Mellan [14] believes that sexual behavior in individuals with SID differ from other people particularly because there is no phase of rational barriers. According to Matulay [11], in general, the more serious ID, the less sexuality. Another argument is that most individuals with SID have seriously defective motor skills or other accompanying disabilities, defective reproductive organs, hormonal disorders – resulting in the possible distortion of sexuality. However, whilst describing characteristics of the sexuality in individuals with moderate and SID, many commonalities can be detected. For these reasons, we have decided to analyze more deeply differences in sexual behavior of individuals with SID in terms of their ID.

3. We assume that individuals of coeducational institutions will show significantly higher frequency and intensity of sexuality in comparison to individuals of non-coeducational institutions. Also, several research studies suggest that many individuals in institutional care have experienced some sexual activities, even though the range of their sexual behavior is probably limited partially due to their sexual repression [12]. This situation can occur in non-coeducational institutions where individuals do not have a possibility, or such possibility is considerably limited, to meet the opposite sex. It can influence the intensity of sexual manifestations as well.

4. It has been suggested that less educated participants, express less positive attitudes to sexuality [2].

5. We studied sexuality in men and in women separately, as sexual expression in women and men is traditionally different and attitudes could be influenced by this factor [2]. According to Weiss and Zvěřina [25] gender has the most significant influence on liberalism and restrictions of sexual attitudes. Despite the fact that many authors discover that attitudes in men and women to sex questions have over the last decades equaled, most research studies testify on a double standard for a male and female sexual role and its reflection in participants' attitudes. Almost all authors agree that attitudes of men are more liberal than attitudes of women in this area.

6. We assume that, the more positive attitudes that employees show in general, the more positive their attitudes to sexuality in individuals with SID will be [2]. Attitudes to sexuality are then often influenced by the personality of experts and that can influence the attitudes to individuals with SID.

7. Study by Cuskelly and Gilmore [2] showed that employees were least positive to questions of parenthood than to other aspects of sex life. We have reached similar conclusions with a questionnaire focused on attitudes of teams of interdisciplinary experts of SCHs' employees, psychologists, psychiatrists, and gynecologists. Most experts did not accept the possibility that individuals with SID have their own children [8].

http://dx.doi.org/10.5772/55782

4. Operationalization of variables

In terms of evaluation of the intensity of the sexual behavior in individuals with SID, we have regarded as sexuality manifestations not only actions of explicitly sexual character, such as intercourse or masturbation, but also dreams, romantic thoughts and temporary or permanent affection, a need to look attractive to a person whom a person with ID grows fond of and whose interest he or she wishes to arouse. To perform data analysis, we have divided sexual behavior manifestations of individuals into two levels:

1. Pregenitality (sexuality does not occur) – asexuality, no interest in any sexual manifestations, manifestations on the level of tactile contacts, caressing, cuddling, touching, snuggling, stroking – rather tactile satisfaction.
2. Sexuality occurs – interest in masturbation, coital contact, etc.

We have made an overview of sexuality manifestations in the first and second level (Table 1).

Code	Sexualitymanifestations ofthefirstlevel	Code	Sexualitymanifestations ofthesecondlevel
K1	Cuddling of same-sex individuals	K5	Touching intimate parts of a same-sex individual
K2	Cuddling of opposite-sex individuals	K6	Touching intimate parts of an opposite-sex individual
K3	Stroking a same-sex individual	K10	Watching erotic pictures and porn
K4	Stroking an opposite-sex individual	K15	Kissing a same-sex individual
K7	Desire to look attractive to a same-sex individual (taking care of one's appearance, highlighting femininity, masculinity)	K16	Kissing an opposite-sex individual
K8	Desire to look attractive to an opposite-sex individual	K17	Indecent public exposure
K9	Listening to love songs, songs about being in love	K18	Sexual assaults (attacks)
K11	Blushing in the presence of the same sex	K19	Intercourse (coitus)
K12	Blushing in the presence of the opposite sex	K20	Masturbation
K13	Individuals rivalry	K21	Fetishes (touching nylons, etc.)
K14	Manifestations of joy, happiness in the presence of the opposite sex		

Table 1. Coding of the chosen sexuality manifestations of the first and second level

Whilst verifying our hypotheses, we monitored differences in specific aspects of sexuality – in terms of their frequency and intensity.

5. Sampling characteristics

A sample group was formed of 459 individuals with ID. In terms of institutional coeducation, 61.1% of individuals lived in coeducational and 15.5% in non-coeducational institutions. In terms of gender, the representation of individuals was relatively equal – 50.3% (221) of individuals were males and 49.7% (218) were females. Out of the total number of individuals, 49.8% (225) of individuals had moderate ID, 31.2% (141) had SID and 1.3% (6) had profound ID. The second sample group included 259 SCHs' employees, 76.5% (198) were females and 16.9% (44) were males, and 6.6% (17) of respondents did not specify the gender. The age was specified in total of 214 employees (45 did not specify it). The youngest respondent was 20, and the oldest was 60 years old. The length of experience was specified in total of 47 employees (212 did not specify it). The average length of experience was 9.17 years; the shortest was one year, the longest 26 years. The length of experience in an institution was specified in total of 219 employees (40 did not specify it). The shortest length of experience was less than one year and the longest was 36 years. Out of the total number of employees, 73.4% (190) were single, 10.4% (27) married, 8.5% (22) stated an option "other" (a widow/divorced) and 7.7% (20) of employees did not specify their family status. In terms of religion, 78.4% (203) of the respondents were religious, 8.1% (21) non-religious, 7.7% (20) nondescript, and 5.8% (15) did not specify it.

6. Research methods

In the course of our research, we used the following methods: direct non-participating observation, questionnaire, and sexual stories – the statements of employees about individuals with SID.

Observational studies. To clarify and describe sexual manifestations, we mainly used direct non-participating observation. Here we focused on developing suitable observation-coding sheets and their records. Observation and the records of sexual manifestations of individuals with SID were enabled by employees volunteering to participate in this study. The first part contained detailed instructions on how to fill it in and data about individuals and institutions. The main part of the observation-coding sheet was formed of: the frequency table of sexual manifestations in SCHs – with the aim to study in details the sexual behavior in SCHs. The employees over a 7-day period observed and marked into an observation sheet in an appropriate column (day and hour) with an assigned symbol/code, observed sexual behavior of individuals in everyday activities during the entire day. The symbols were listed in a separate appendix and divided into individuals-to-individuals sexual behavior and individuals-to-employees sexual behavior. If an observed manifestation was not recorded in the table, the respondents were instructed to complete it and mark its frequency in the table (on appropriate days and hours).

1. A Table listing sexual manifestations of an SCH client – with the aim to identify consequences of coded sexual behavior (reactions of employees and subsequent reactions of

individuals). The employees were supposed to characterize more closely the specific observed sexual behavior of a client and their eventual reactions in the provided table.

2. The overall characteristics sheet of sexual behavior of a client – the purpose of the third part of the observation-coding sheet was to obtain materials to create sexual stories or statements on sexual behavior of individuals. The employees were supposed to give characteristics of sexual behavior of individuals in this part, based on long-term observations of a client.

7. Attitudes to Sexuality Questionnaire (ASQ)

Given the fact that there are rather unfavorable attitudes in public to sexuality in individuals with ID, we were interested in SCHs employees' attitudes. To study these, we used a standardized ASQ questionnaire – Attitudes to sexuality questionnaire (individuals with an ID) by Cuskelly and Gilmore [2]. ASQ consists of two parts:

1. Attitudes to sexuality questionnaire (Individuals from the general public: ASQ – GP).
2. Attitudes to sexuality questionnaire (Individuals with ID: ASQ – ID) – This part contains 34 questions divided into four aspects: sexual rights, parenthood, non-reproductive sexual behavior, and self-control.

Respondents answered using a 6-point scale, varying from "strongly agree" to "strongly disagree". Both questionnaires contain questions concerning sexuality in women and sexuality in men, answered by respondents separately. The advantage of the questionnaire is a wide range of areas, which are available to select for research. The disadvantage could be its length that could discourage potential candidates.

8. The employees' opinions on the sexuality of individuals with SID and sex education in SCHs questionnaire

The questionnaire was designed with the aim to obtain data concerning sexuality and sex education in SCHs. The analysis was relatively a time-consuming process that went through several stages and relied on various sources of data. Thus, we obtained a holistic view of factors that can influence result. Once the questionnaire had been reviewed, a two-phase pilot study began with the aim of confirming whether respondents understood questions and were able to answer them without problems. The questionnaire items contain categories and phenomena elicited within pre-research. The questionnaire monitored the following research areas:

1. Quality indicators of social services provided by a particular institution (a size of an institution, coeducation, a number of individuals sharing a room, education of employees, etc).
2. Sexuality and its manifestations in individuals with SID in SCHs.

3. Training and awareness of employees and individuals in SCHs in terms of sexuality and sex education of individuals with SID.

4. Opinions of employees on sexuality and sex education in individuals with SID and the current situation in sex education implementation in SCHs.

5. Problem in sexuality and sex education and ways of solving them in a particular institution.

The questionnaire, that was designed for this research consisted of 46 items. Despite the fact that we were aware of somewhat broad questionnaire, we were compelled to use the items in such quantity in regard to absence of valid data to build on, and rather broad-spectrum aims that we had set. Respondents were instructed via a letter containing essential instructions to fill in the questionnaire. The questionnaire items were mostly structured in such a manner, that respondents were offered a certain number of predefined answer choices. The questionnaire consists of mostly semi-closed items and respondents were able to fill in the offered answers and their own opinions in an option *other*. Besides answer choices, there were also unstructured items in the questionnaire to allow respondents formulate their own answers. It should be noted that in the Slovak Republic, such research of sexual behavior of individuals with SID and the implementation of sex education in SCHs has been generally missing.

9. The statements — Written sexual stories of SCHs' employees and individuals with SID

To be complete, the information was acquired not only from respondents' point of view through the questionnaire, but primarily from providers and users of social services. This part was carried out through interviews, using a recorder with a transcription following.

10. Research results and interpretation

10.1. The first procedure results: an analysis of the sexuality of individuals with SID

In regard to outlined disagreement of experts and with the aim to describe the sexuality of individuals with SID in SCHs and to draw conclusions, we will interpret the results using observation-coding sheets.

The first phase of analysis: the general interpretation of the frequency of sexuality manifestations occurrence in individuals with SID

We focused on the general interpretation of the frequency of sexuality manifestations occurrence in 459 individuals in SCHs. There were in total 3457 sexual manifestations of individuals-to-individuals with SID registered. From the general evaluation of the observation coding sheets we found that there were 68.8% (2273) of sexuality manifestations of the first level (so-

http://dx.doi.org/10.5772/55782

called pregenitality) in the following order: manifestations of joy, happiness in the presence of the opposite sex (12%); cuddling of same-sex individuals (11.2%); stroking of a same-sex individual (10.5%); stroking of an opposite-sex individual (9.2%); cuddling of opposite-sex individuals (7.3%); listening to love songs, songs about being in love (5.8%); other manifestations of this level were registered with an occurrence of less than 5% (Figure 1).

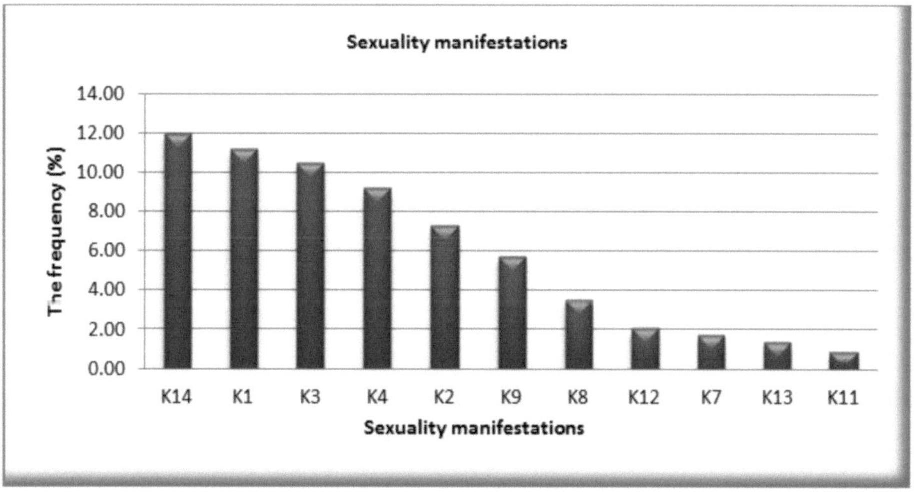

N – the number of sexual manifestations, K1 – cuddling of same-sex individuals, K2 – cuddling of opposite-sex individuals, K3 – stroking of a same-sex individual, K4 – stroking of an opposite-sex individual, K7 – desire to look attractive to a same-sex individual (taking care of one's appearance, highlighting femininity, masculinity), K8 – desire to look attractive to an opposite-sex individual, K9 – listening to love songs, songs about being in love, K11 – blushing in the presence of the same sex, K12 – blushing in the presence of the opposite sex, K13 – individuals rivalry, K14 – manifestations of joy, happiness in the presence of the opposite sex. The results reflect representation of the first level manifestations in individuals with SID.

Figure 1. The frequency of the first level sexuality manifestations in individuals with SID toward other individuals.

Some of the registered sexuality manifestations of the first level are demonstrated through statements of SCHs' employees:

A 34-year-old female with SID. *The client likes to groom herself in front of a mirror. She blushes when an opposite-sex client comes. During the day she cuddled his waist more times. She listens to love songs and smiles while doing it. In the evening she got changed more times and watched herself in a mirror again, checking her look.*

A 39-year-old male with moderate ID. *The client listens to love songs. He expresses a lot of joy while doing it. He is happy when they speak in the room about a female client he fancies. He cares about his appearance and he wants to look attractive. He beautifies himself, puts some cream on and combs his hair. He repeats the same ritual several times a day. He smiles and he's nice, he likes the girls from the next room.*

In regard to the sexuality manifestations of the second level (in total 980, i.e. 28.4% of all manifestations), they primarily consist of masturbation (8.8%), kissing an opposite-sex individual (6.4%), kissing a same-sex individual (2.8%), touching intimate parts of a same-sex individual (2.2%), intercourse (coitus) (1.3%) (Figure 2).

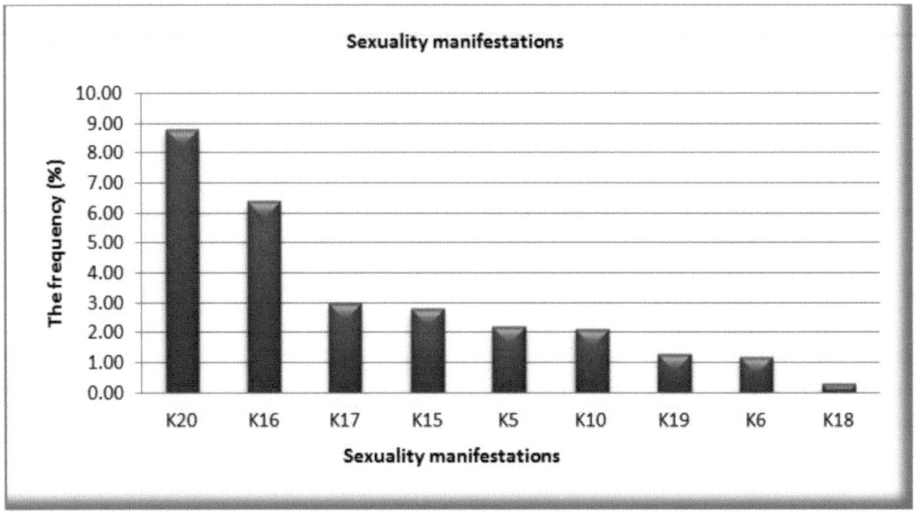

N – number of sexual manifestations, K5 – touching intimate parts of a same-sex individual, K6 – touching intimate parts of an opposite-sex individual, K10 – watching erotic pictures and porn, K15 – kissing a same-sex individual, K16 – kissing an opposite-sex individual, K17 – indecent exposure, K18 – sexual assaults, K19 – intercourse, K20 – masturbation. The results reflect that besides the first level manifestations there are also manifestations of the second level in individuals with SID. These data show that sexuality occurs in individuals with SID and that individuals with SID are also interested in activities of sexual character, such as masturbation or coitus.

Figure 2. The frequency of the second level sexuality manifestations in individuals with SID toward other individuals.

The analysis of the frequency of sexuality manifestations of the first and second level does not include other sexuality manifestation (204, i.e. 5.9% of all recorded manifestations), which was provided by the employees, but without further specification of particular manifestations, therefore, we were not able to specify a relevant level. The results of sexual manifestations of the second level are supported by statements of SCHs' employees:

A 44-year-old individual with SID. Besides the stated diagnoses the client has a motor disability, but nonetheless her sexual activity is increased. It has led to coitus with an opposite-sex client; in same-sex individuals it is manifested through cuddling, kissing, etc.

A 31-year-old individual, with moderate IS. *The client manifests his sexuality particularly through masturbation on a bed lying flat on his stomach and rubbing on a bed forward and backward (on average twice or three times a day – while taking a rest, after lunch and before bedtime). He behaves as if he needs privacy and dark – he doesn't want anybody to take any notice of him. He fancies uncovered female toes, gazes at them and tries to touch them. In the past, he caught a foot of a new educator who had her toes uncovered and he tried to rub it against his penis. His sexual potency is increased in summer, he's moody and it lasts two or three weeks.*

An 18-years-old individual with Down syndrome, with moderate ID.*There are mostly non-verbal manifestations; it's mostly difficult to understand his words. He often throws himself at female individuals and tries to kiss them. He's too vivid and hyperactive. He masturbates every day without ejaculating and if it takes him too long, he gets nervous and hits his penis.*

On the basis of our results, we can state that in SCHs' individuals, a sexual need is present and is, in regard to ID and the conditions in which individuals live manifested through the above-mentioned sexual manifestations. Despite that, it is obvious from some of the employees' statements that they classify sexual activities as an asexual manifestation. A very interesting factor that would enable a deeper analysis of sexuality, is perhaps an environment in which individuals live (coeducation, a number of individuals sharing one room, etc.), as well as opportunities to meet each other, be alone, attitudes of employees to individuals' sexuality, etc.

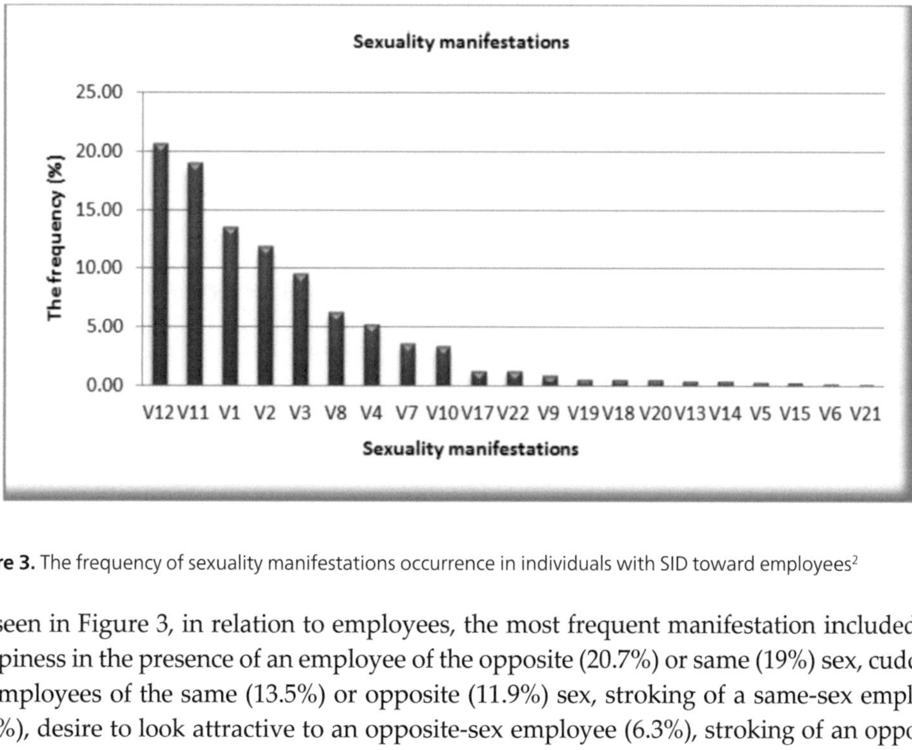

Figure 3. The frequency of sexuality manifestations occurrence in individuals with SID toward employees[2]

As seen in Figure 3, in relation to employees, the most frequent manifestation included joy, happiness in the presence of an employee of the opposite (20.7%) or same (19%) sex, cuddling of employees of the same (13.5%) or opposite (11.9%) sex, stroking of a same-sex employee (9.6%), desire to look attractive to an opposite-sex employee (6.3%), stroking of an opposite-sex employee (5.2%). Other manifestations were registered with an occurrence of less than 5%. From the listed overview of the registered sexuality manifestations we draw attention to non-selective sexual behavior of individuals, which is many times focused on employees of the opposite but also the same sex. Here, we present a few examples, documenting this behavior:

A 22-year-old individual with moderate ID.*When larking and being relaxed, looking at a female educator the client gets ecstatic and reddens. The intensity grows until the moment of physical touch, when a male educator has to step in.*

A 35-year-old person with SID. *The client likes listening to love songs. She's very happy seeing an employee starting his shift. She blushes when a female social worker comes. She beautifies herself, puts some cream on and combs her hair. She repeats the same ritual.*

We found reasons of non-selective sexual behavior in outer (absence of sex education, providing unsuitable models, institutionalized environment, etc.) and inner conditions of individuals with SID (a primary deficit – the total subnormal intelligence, secondary deficits – limited thinking, emotional immaturity, non-independence, lack of experience, non-criticality, suggestibility, inability to differentiate a level of behavioral adequacy, etc.). For these reasons, there is a need to explain to individuals with SID sexual behavior in context, e.g. we do not cuddle all people and we do not cuddle them without a reason. It is obvious, that clear instructions and counseling on how to solve practical problems manifested in relation to individuals' sexual behavior, are primarily necessary for SCH' employees.

10.2. The second phase of analysis: factors influencing the sexuality of individuals with SID

Human sexual behavior is primarily determined by biological dispositions. The outer environmental circumstances could significantly modify its particular manifestations. In the second phase of our study, we focused on a finding, whether the sexuality of individuals with SID are influenced by endogenous factors, such as sex and a level of ID. At the same time we detected the influence of an endogenous factor, i.e. the type of the institution in terms of its coeducation on sexuality in people with SID.

10.3. A comparison of the sexuality of individuals in terms of gender

In the first hypothesis we were interested whether men with SID would show a significantly higher frequency and intensity of the sexuality in comparison to women with SID. To analyze the nature of research data we used the Anderson-Darling Normality Test, together with a calculation of the basic descriptive characteristics of the samples.

There were in total 221 males and 218 females compared. None of the research samples showed a normal distribution ($p < 0.05$) and therefore we used the non-parametric Mann-Whitney U-test to compare the samples.

Numberofmanifestations(intotal)	N(thenumberofsexualmanifestations)	Median
Men	221	5.000
Women	218	9.000
Point estimate for ETA1-ETA2 is -2.000 95.0 Percent CI for ETA1-ETA2 is (-4.000; -1.000) W = 44638.5 Test of ETA1 = ETA2 vs. ETA1 not = ETA2 is significant at 0.0027 The test is significant at 0.0027 (p-value) (adjusted for ties)		

Table 2. The Mann-Whitney U-test coefficients to compare the frequency of sexual manifestations of individuals in terms of gender

Table 2 shows that in our sample, using the Mann-Whitney U-test, the p-value[3] was 0.0027, suggesting that there was a significant difference in the sexuality manifestations between men and women with SID.

Next, we observed differences in the intensity of the sexuality in men and women with SID. To analyze our data, we similarly used the Anderson-Darling Normality Test, together with a calculation of the basic descriptive characteristics of the samples. To compare the samples we used Mann-Whitney U-test. As shown in Tables 3 and 4, there were not significant differences in the intensity of sexuality manifestations between men and women with SID.

Manifestationsofthefirstlevel(intotal)	N(thenumberofsexualmanifestations)	Median
Women	143	6.000
Men	105	6.000
Point estimate for ETA1-ETA2 is -0.000 95.0 Percent CI for ETA1-ETA2 is (-1.001; 1.000) W = 17623.5 Test of ETA1 = ETA2 vs. ETA1 not = ETA2 is significant at 0.7478		

Table 3. The Mann-Whitney U-test coefficients to compare the first level sexuality manifestations of individuals in terms of gender.

Manifestationsofthesecondlevel(intotal)	N(thenumberofsexualmanifestations)	Median
Women	70	3.500
Men	99	4.000
Point estimate for ETA1-ETA2 is -1.000 95.0 Percent CI for ETA1-ETA2 is (-1.000; -0.001) W = 5572.0 Test of ETA1 = ETA2 vs. ETA1 not = ETA2 is significant at 0.2283		

Table 4. The Mann-Whitney U-test coefficients to compare the second level sexuality manifestations of individuals in terms of gender.

Based on these results, we suggest that our first hypothesis, in which we assumed men with SID would show significantly higher frequency and intensity of sexuality in comparison to women with SID, has not been confirmed. In this respect, a statement of Weiss and Zvěřina [25] has been proved, suggesting that there is only little evidence that differences in a sexual activity between the genders would be conditioned biologically. According to these, social norms in women prevail regarding a sex and emotions relationship. To obtain a deeper qualitative analysis of the data and as we wanted to obtain the maximum amount of information on differences in sexuality between men and women with SID, we proceeded with the next part studying the differences in the frequency of occurrence of particular sexual manifestations, which were common, and predicted significant differences in terms of their representation in men and women with SID.

Manifestation'scode	Men	Women	Sum	Wstatistics	p-value
K15	61	36	97	241	0.0021*
K9	111	86	197	1055.5	0.0013*
V1	13	148	161	419	0.0074*
V11	16	249	265	244	0.0008*
V12	218	58	276	519.5	0.0012*

K15 – kissing a same-sex individual, K9 – listening to love songs, songs about being in love, V1 – cuddling of same-sex employees, manifestations of joy, happiness in the presence of a same-sex employee, V12 – manifestations of joy, happiness in the presence of an opposite-sex employee,*p < 0.05 – a statistically significant difference

Table 5. Significant effects of gender on the frequency of sexuality manifestations.

As shown in Table 5, in women, there were significantly more manifestations on the level of cuddling and signs of joy and happiness, especially in relation to employees than in men. Men showed more manifestations, such as listening to love songs, songs about being in love, kissing a same-sex individual and manifestations of joy, happiness in the presence of an opposite-sex employee than women.

10.4. A comparison of sexuality in terms of ID levels

In the second hypothesis we asked whether individuals with a moderate ID would show significantly higher frequency and intensity of the sexuality in comparison to individuals with a SID. We analyzed the compared groups using the Anderson-Darling Normality Test, together with a calculation of the basic descriptive characteristics of the samples (225 individuals with a moderate ID and 141 individuals with a SID.

Numberofmanifestations(intotal)	N	Median
MID	225	8.000
SID	141	4.000

Point estimate for ETA1-ETA2 is 3.000

95.0 Percent CI for ETA1-ETA2 is (2.001; 5.000) W = 45584.0

Test of ETA1 = ETA2 vs. ETA1 not = ETA2 is significant at 0.0000

The test is significant at 0.0000 (p-value) (adjusted for ties)

MID – moderate ID, SID – severe intellectual disability, N – number of sexual manifestations

Table 6. The Mann-Whitney U-test coefficients to compare sexuality manifestations frequency of individuals in terms of ID level.

Based on our results (Table 6), there was a significant difference in the sexual manifestations between individuals with moderate and severe ID.

http://dx.doi.org/10.5772/55782

To study the differences in the intensity of the sexuality in terms of an ID level, we used Anderson-Darling Normality Test, together with a calculation of the basic descriptive characteristics of the samples. As the following calculations of the samples comparison specified in Tables 7 and 8 show, we did not detect significant differences in the sexual manifestations intensity between individuals with moderate and SID.

Manifestationsofthefirstlevel(intotal)	N	Median
MID	149	6.000
SID	48	4.000
Point estimate for ETA1-ETA2 is 1.000 95.0 Percent CI for ETA1-ETA2 is (-0.001; 2.999) W = 15364.5 Test of ETA1 = ETA2 vs. ETA1 not = ETA2 is significant at 0.0744		
MID – moderate intellectual disability, SID – severe intellectual disability, N – number of sexual manifestations		

Table 7. The Mann-Whitney U-test coefficients to compare the first level sexuality manifestations of individuals in terms of ID level.

Manifestationsofthesecondlevel(intotal)	N	Median
MID	79	3.000
SID	52	4.000
Point estimate for ETA1-ETA2 is 0.000 95.0 Percent CI for ETA1-ETA2 is (-0.999; 1.001) W = 5127.5 Test of ETA1 = ETA2 vs. ETA1 not = ETA2 is significant at 0.6858		
MID – moderate intellectual disability, SID – severe intellectual disability, N – number of sexual manifestations		

Table 8. The Mann-Whitney U-test coefficients to compare the second level sexuality manifestations of individuals in terms of ID level.

On the basis of these results, we suggest that the second hypothesis, in which we assumed that individuals with moderate ID would show a significantly higher frequency and intensity of the sexuality in comparison to individuals with SID, has not been confirmed. Whilst comparing differences in an occurrence of the specific sexuality manifestations, we found that in individuals with moderate ID prevailed sexual behavior manifestations on the level of pregenitality, whereas in individuals with SID were more than in individuals with moderate ID registered activities of sexual character. We consider a difference in the frequency of masturbation occurrence interesting, as it occurred more often in individuals with SID than in individuals with moderate ID. In individuals with SID we found out at the same time a higher frequency of indecent exposure and intercourse occurrence. We can conclude that it was due to the fact that individuals with SID expressed the activities more often in front of the staff than in privacy. However, the differences in the specific sexuality manifestations were statistically not significant.

10.5. A comparison of the sexuality in terms of institution's coeducation

We used the Anderson-Darling Normality Test together with a calculation of the basic descriptive characteristics of the samples. We compared sexual manifestations of 298 individuals of coeducational institutions and 68 individuals of non-coeducational institutions.

Numberofmanifestations(intotal)	N(thenumberofsexualmanifestations)	Median
A coeducational institution	298	4.000
A non-coeducational institution	68	10.000
Point estimate for ETA1-ETA2 is -4.000 95.0 Percent CI for ETA1-ETA2 is (-5.000; -3.000) W = 50738.0 Test of ETA1 = ETA2 vs. ETA1 not = ETA2 is significant at 0.0000 The test is significant at 0.0000 (p-value) (adjusted for ties)		

Table 9. The Mann-Whitney U-test coefficients to compare sexuality manifestations frequency of individuals of coeducational and non-coeducational institutions.

As we show in Table 9, $p < 0.05$, therefore we may state that there was a significant difference in the frequency of sexual manifestations between individuals living in non-coeducational or coeducational institutions.

It is also possible to conclude from the box graph, that in the coeducational institutions there was an excessive occurrence of sexually active individuals registered, in a non-coeducational institution such individuals did not occur.

Next we studied the differences in the intensity of sexuality in terms of institution's coeducation. Not even in this case the samples showed a normal distribution. Calculations of the non-parametric Mann-Whitney U-test presented in Tables 10 and 11 show, that there were not significant differences in intensity of sexual manifestations in individuals with SID in terms of institution's coeducation ($p > 0.05$).

Manifestationsofthefirstlevel(intotal)	N(thenumberofsexualmanifestations)	Median
A coeducational institution	152	5.000
A non-coeducational institution	49	4.000
Point estimate for ETA1-ETA2 is 1.000 95.0 Percent CI for ETA1-ETA2 is (-0.000; 2.000) W = 15766.0 Test of ETA1 = ETA2 vs. ETA1 not = ETA2 is significant at 0.2429		

Table 10. The Mann-Whitney U-test coefficients to compare the first level sexuality manifestations of individuals of coeducational and non-coeducational institutions.

On the basis of these results, we suggest that the third hypothesis, in which we assumed that individuals of coeducational institutions would show a significantly higher frequency and

http://dx.doi.org/10.5772/55782

Manifestationsofthesecondlevel(intotal)	N(thenumberofsexualmanifestations)	Median
A coeducational institution	119	3.000
A non-coeducational institution	12	2.000
Point estimate for ETA1-ETA2 is 1.000 95.1 Percent CI for ETA1-ETA2 is (-1.000; 2.000) W = 7981.0 Test of ETA1 = ETA2 vs. ETA1 not = ETA2 is significant at 0.3128		

Table 11. The Mann-Whitney U-test coefficients to compare the second level sexuality manifestations of individuals of coeducational and non-coeducational institutions.

intensity of the sexuality in comparison to individuals of non-coeducational institutions, has not been confirmed. Further structural analysis of the sexuality enabled us to search for an answer to a question, which of the observed sexual activities are specific for the institution or what are differences in specific sexual behavior of individuals of coeducational and non-coeducational institutions.

Next, we evaluated differences in sexuality manifestations of individuals of coeducational and non-coeducational institutions. The results of the analysis and the statistical difference between the specific sexual manifestations of individuals of coeducational and non-coeducational institutions point to the main conclusions presented in Table 12.

Manifestation'scode	CI	NI	Sum	Wstatistics	p-value
K2	121	1	122	-	-
K4	138	4	142	-	-
K5	47	3	50	-	-
K6	20	-	20	-	-
K10	26	-	26	-	-
K16	90	-	90	-	-
K19	26	-	26	-	-
K20	193	5	198	-	-
V11	74	138	212	921.5	0.0141*

K2 – cuddling of opposite-sex individuals, K4 – stroking an opposite-sex individual, K5 – touching intimate parts of a same-sex individual, K6 – touching intimate parts of an opposite-sex individual, K10 – watching erotic pictures and porn, K16 – kissing an opposite-sex individual, K19 – intercourse (coitus), K20 – masturbation, V11 – manifestations of joy, happiness in the presence of a same-sex employee, CI – coeducational institution, NI – non-coeducational institution, * $p < 0.05$ – a statistically significant difference

Table 12. Differences in the frequency of sexuality manifestations occurrence of individuals in terms of institution's coeducation.

As shown in Table 12, a statistically significant difference in the frequency of sexuality manifestations was confirmed only in one manifestation, i.e. manifestations of joy, happiness in the presence of a same-sex employee – significantly more in non-coeducational institutions than in coeducational institutions. Despite what has been stated, we want to highlight other differences, which we consider very significant in this regard. In terms of the first level sexuality manifestations, there were in the coeducational institutions more manifestations focused on opposite-sex individuals. Whilst observing the frequency of sexuality manifestations occurrence of the second level, we highlight a significant difference in implementation of masturbation, registered in the individuals of the coeducational institutions more often than in the individuals of the non-coeducational institutions, and in implementation of an intercourse, similarly registered more often in the coeducational than in the non-coeducational institutions. In non-coeducational institutions, as shown by other studies, sexual behavior on the level of masturbation or an intercourse is probably limited, partially due to sexual manifestations repression of the individuals.

11. The second procedure results: an analysis of SCHs employees' attitudes to sexuality and sex education in individuals with SID

11.1. The first phase of interpretation: the interpretation of employees' attitudes

Here we focused on the interpretation of employees' attitudes to sexuality in individuals with SID. The complex analysis including all the sexuality concludes an ambiguous evaluation of sexuality and sex education in individuals with SID by SCHs' employees. This suggests that attitudes of employees to sexuality in individuals with SID are neutral.

In the next part, we present results of the interpretation of employees' attitudes to particular sexuality aspects in individuals with SID, which are included in our attitudes questionnaire.

	Thesexualityaspects	**Scale** **1-** 3 **- 6**
		P——— N
1.	Sexual rights of individuals with more severe ID	**3.5385***
		P——— N
2	Parenthood of individuals with more severe ID	**4.0714***
		P——— N
3	Non-reproductive sexual behavior of individuals with more severe ID	**3.1000***
		P——— N
4.	Self-control of individuals with more severe ID	**3.5000***

P – positive attitude, N – negative attitude, * – the median values

Table 13. An overview of employees' attitudes to the sexuality.

http://dx.doi.org/10.5772/55782

As shown in Table 13, positive attitudes of the interviewed staff occurred only in the area of non-reproductive sexual behavior. We introduce some examples of the respondents' most positive: Masturbation in privacy is an accepted form of a sexual manifestation in men/women with ID. It is sensible to maintain privacy for men/women with ID, who want to masturbate at home. We encounter a neutral attitude to the questions of self-control.

11.2. The second phase of interpretation: The analysis of the influence of specific factors on employees' attitudes

A comparison of attitudes in SCHs' employees to sexuality in individuals with SID in regard to training programs.

According to our fourth hypothesis, employees who completed a training program in sex education would show significantly more positive attitudes to sexuality in individuals with SID in comparison to employees who did not complete training programs in sex education.

Trainingprogram	N(thenumberofsexualmanifestations)	Median
Yes	66	3.5714
No	193	3.4464
	Point estimate for ETA1-ETA2 is 0.0714 95.0 Percent CI for ETA1-ETA2 is (-0.0536; 0.1785) W = 9198.0 Test of ETA1 = ETA2 vs. ETA1 not = ETA2 is significant at 0.2398	

Table 14. The Mann-Whitney U-test coefficients to compare a training program in sex education.

The calculated p-value seen in Table 14 is not significant; therefore we suggest that there was not a significant difference in attitudes between employees who have completed a training program in sex education, and those who have not.

According to our hypothesis, employees who completed a training program in sex education would show significantly more positive attitudes to sexuality in individuals with SID in comparison to employees who did not complete a training program in sex education was not confirmed. The presented conclusion is inconsistent with the research findings presented by Moatti, Manesse and colleagues [25] who believed that the more restrictive attitudes are expressed by less erudite people. We may suppose, that the attitudes of the employees, who have completed a training program, could have been influenced by a level and range of a program itself and therefore their cognitive part of attitudes was not positively influenced. This could also be a result of the respondents' negative experiences with sexuality in individuals with SID, resulting from everyday direct activities with them. Hence we find the results to be a strong argument for improving the level of training programs of employees in sex education in individuals with SID.

The comparison of attitudes in employees to sexuality in individuals with SID in terms of gender of employees

In the fifth hypothesis, we assumed that attitudes of male employees would be significantly more positive to sexuality in both male and female individuals with SID in comparison to attitudes of female employees. We found that gender does not have a statistically significant influence on attitudes of employees. There are several studies that suggest that gender itself has the most significant influence on liberalism or restriction of sexual attitudes. As we mentioned before in our research study, almost all authors agree, that attitudes of men are in this area more liberal than attitudes of women. According to Weiss and Zvěřina [25], however, many authors discover that attitudes of men and women to sexual questions have over the last decades come closer. Nowadays, men and women become more alike not only in work life, but as Walker-Hirsch [24] writes, even in sexual expression and in attitudes to sexuality – and our results support it. It could be caused by the general social climate, too, which becomes more liberal (influence of mass media, etc.) and allows women to freely express their opinions and adopt a point of view to them.

A comparison of attitudes in SCHs' employees to sexuality in individuals without a disability versus individuals with SID

The attitudes to sexuality in individuals with ID are certainly linked to our attitudes to human sexuality in general. Hence, in the sixth hypothesis, we assumed that the more positive attitudes would employees show to sexuality in individuals without a disability, the more positive their attitudes to sexuality in individuals with SID would be. To test this hypothesis we used a correlation analysis. At the beginning of the correlation analysis we picture the researched data graphically using a two-dimensional point graph, where each point represents one pair of measurements.

It is possible to estimate positive correlation between attitudes to sexuality in individuals without a disability and individuals with SID. We further verified this assumption using the Pearson correlation coefficient (r)[4] (r = 0.6085; Table 15). Our data suggest that those attitudes to sexuality in individuals with no SID significantly correlate with sexuality in individuals with SID.

N=259	Correlation–significanceonthelevelα<0.0500						
	AM	SD	*r*(X,*Y*)	*r*^2	*t*	*p*	N
An attitude to sexuality in individuals without a disability	2.7473	0.6229	-	-	-	-	-
An attitude to sexuality in individuals with a more severe ID	3.5240	0.4406	0.6086	0.3703	0.0000	0.0000	259

AM – the arithmetic mean, SD – standard deviation, r – the Pearson correlation coefficient, XY – the observed variables, t, p – coefficients and values calculated by the test, N – number of respondents

Table 15. A table of auxiliary calculations of the Pearson correlation coefficient of an attitudes correlation to sexuality in individuals without a disability and individuals with more severe ID.

On the basis of the presented data in Table 15 we suggest that the sixth hypothesis, in which we assumed that the more positive attitudes in employees to sexuality in the general public, the more positive their attitudes to sexuality in individuals with SID would be, has been confirmed. Thus, the assumption has been confirmed that the way that we perceive sexuality in individuals with SID is to a great extent influenced by our personal experience or an attitude to own sexuality. Furthermore, an opinion of SCHs' employees on sexuality in these individuals is influenced by an attitude to human sexuality in general. The assumption of sex education in these individuals is a possibility to obtain personal positive opinions, attitudes, and experiences to human sexuality. In the seventh hypothesis we assumed, that employees would show significantly more positive attitudes to non-reproductive sexual behavior in comparison to parenthood of individuals with SID.

Sexuality	N(thenumberofsexualmanifestations)	Median
Non-reproductive sexual behavior	259	3.1000
Parenthood	259	4.0714
	Point estimate for ETA1-ETA2 is -0.9571 95.0 Percent CI for ETA1-ETA2 is (-1.0571;-0.8572)W = 43793.5 Test of ETA1 = ETA2 vs. ETA1 not = ETA2 is significant at 0.0000 The test is significant at 0.0000 (p-value) (adjusted for ties)	

Table 16. The Mann-Whitney U-test coefficients to compare attitudes to non-reproductive sexual behavior and parenthood.

Using The Mann-Whitney U-test, the p value was less than 0.05 indicating that there was a statistically significant difference between the two groups (Table 16). Based on our data, we can suggest that employees had a significantly more positive attitude to non-reproductive sexual behavior than to parenthood in individuals with SID.

On the basis of the above results, we form a thesis that the seventh hypothesis, in which we predicted that employees would show significantly more positive attitudes to non-reproductive sexual behavior in comparison to parenthood in individuals with SID, has been confirmed. According to this hypothesis, employees (and a society in general) incline more to sexual activities in individuals with SID, which do not pose a pregnancy risk. Moreover, prejudices persist in a society as well as in providers of social care resulting from lack of information, and that could, as a result, lead to the denial of basic rights of individuals with SID – the right to have a child. It is a very sensitive issue – on one hand the right to have a child cannot be violated because it is an individual with SID, on the other hand there is a risk of immense emotional value in relation to a future child and future parents.

Finally, it should be noted that attitudes are the key factors, which explain human social behavior. We believe that the results of our research on this issue would create an opportunity to study the quality of educational impact of employees on individuals with SID in terms of their sexuality.

11.3. The third procedure results: An analysis of sexuality in individuals with SID and sex education implementation in SCHs

In Slovakia, sexuality in individuals with ID is only marginally studied. Yet, sexual partnerships are fundamental elements of a social life and have important social consequences [9]. To what extent are conditions for freedom of sexual expression in individuals with SID and the quality of sex education implementation in SCHs fulfilled, is beyond the scope of this study. Our research study is the first of its kind in Slovakia.

The quality indicators of social services provided by an institution

The first objective in this phase of research was to identify quality indicators of social services provided by institutions in general. We were particularly interested in the size of the institution, the number of individuals sharing each room, and the number of staff. These indicators would certainly influence sex education in these institutions. The largest institution participating in this research provides its services to 240 individuals with multiple disabilities in people 18 years and older. The smallest institution participated in our research provides its services to 18 individuals with ID from birth to 18 years old. Of SCHs that participated in research, 82.1% (23) of institutions provide their services to more than 30 individuals. We used the limit of 30 individuals, as this limit in particular is considered by European experts to be the key in terms of quality of service assessment. According to these experts, in an institution with over 30 residents, there is higher probability, that the basic human rights will be violated [7]. An important quality indicator of provided residential services is also the number of individuals sharing a room. We investigated the minimum and maximum number of individuals sharing a room. Around 25% (7) of the questioned institutions reported a minimum of one resident per room. On the other hand, rather serious information was that one institution stated a minimum of 8 residents living in one room. Surprisingly, in one case the maximum number was 11 individuals. Out of 28 SCHs, only one institution was non-coeducational. We compared sexual manifestations of 298 individuals placed in coeducational institutions and 68 individuals placed in non-coeducational institutions.

On the basis of our research, it is possible to suggest that there are relatively big differences in the institutions in terms of quality of services provided. On one hand, the Slovak Republic used modern facilities offering their services to individuals who have their own room and on the other hand, there are big, sometimes even non-coeducational institutions where individuals have to share with ten other individuals. We believe that an institution should open up as much as possible to an ambient society, to use social teaching of individuals directly "in terrain" to the limit. Being in touch with the general population motivates individuals to constantly develop and learn. Thanks to a very important feedback, they learn faster, and a positive feedback motives them much more than learning in a non-stimulating rigid environment, which does not allow an individual to put everything learnt into practice.

Partner's sexual activities and their conditions in SCHs

Circumstances of partners and sex life are certainly one of the basic indicators of human sexual behavior. These manifestations could have an influence on sexual behavior of an individual with SID. Therefore, we asked questions concerning sexual rights, the sexuality, conditions of forming partnerships, and satisfying of individuals' sexual desire in SCHs. We also included questions in the questionnaire that detected attitudes of respondents on application of sexual rights of individuals with SID. We found that although most employees were able to accept right to sex and partnership in individuals with SID, they express negative attitudes to the right of these people to marriage. The respondents were even stronger in their disapproval of the right to parenthood in individuals with SID. It confirms the findings acquired from the attitudes questionnaire, where the most negative attitudes had the respondents in particular to the questions of parenthood in individuals with SID. Here we provide an example of a client's interest in being in a marriage according to a statement of a social care worker:

A 46-year-old female with moderate ID.*The client was telling me about her son's father. She likes talking about men. She asks various questions. She often asks about her music therapy teacher. Today she was telling me that somebody will marry her. She does not like to hear that a man, whom she had a son with, has got married and has children. She longs for a family life with her dream husband. She likes to put some make-up on in order to be attractive for her music therapy teacher.*

In the next questionnaire item, we aimed to study whether employees encountered sexual manifestations in individuals with SID. We found that almost half (52.1%) of employees have encountered such manifestations. Similarly, almost one third (28.9%) of the respondents expressed negative attitudes toward this. The respondents were also supposed to specify the type of sexual behavior of individuals. The most commonly answer suggests a preference for masturbation over other sexual manifestations (32%). Other manifestations included: kissing (17.8%), cuddling (13.1%), touching intimate parts of another client (9.7%), stroking, caressing (7.7%) and an intercourse (5%). We consider these results to be another evidence of the presence of sexuality in individuals with SID and an interest in activities of sexual character (masturbation, coitus). We also consider it as a strong argument for sex education implementation in these individuals. Occurrence of masturbation is demonstrated here through statements of an SCH educator:

A 42-year-old female, with moderate ID.*The client puts her hands between her legs and touches her sex organs at least once a day, either in the evening in her bed or in the room on the floor. She sways and lets out different sounds. She's very irritated. She sometimes goes to an empty room and repeats her behavior, she often screams furiously.*

Our results support the thesis of Tóthová [23] about insufficient respect of sexual needs and sexual rights of SCHs' individuals. We have also come to the same conclusion, that most institutions do not create conditions for implementation of sex and partners life of individuals with SID. We evaluate it negatively in terms of violation of sexual rights of such individuals to the freedom to pursue their own sexuality. Only less than one-third of the employees stated, that individuals did have such an opportunity, but only on the premises of SCHs. Unfortunately, only 5% of the questioned institutions allowed to create sexual relationships and partnerships even outside of institution. We give an example of a client's interest in being in a partnership with a female client:

A 36-year-old male with SID. *The client often holds a hand of an opposite-sex client. He's already told me couple of times, that he'd want to be in a room just with her. They're walking together in the corridor and they're happy. The whole morning they've been seating together, holding their hands* *and sometimes they're talking as well. He was passive on that day, she was active and was looking for him. When she managed to find him, he was lying in his bed. She was sitting at his side, smiling and was content. Then she took his hand and they went for a walk again. After lunch she took some meat of her plate and went to give it to him. In the evening they shared food again. I've registered deep manifestations of mutual affection. Then they started to kiss, he was stroking her. She attracts his attention, e.g. I've lost weight, I've washed.*

Those who answered that their institutions created conditions for partners and sex life of individuals with SID were then asked to specify its concrete form. We found that in SCHs, a couple is mainly allowed to meet up during common activities at the institution. The results show that in SCHs conditions are not created to allow individuals with SID to pursue their own sexuality. Also currently, such form of cohabitation collides perhaps with prejudice and entrenched ideas of asexuality of these people.

Another question investigated the opinions of employees on enabling implementation of sexual life in individuals with SID. Here, implementation of sexual life refers to enabling sexual life according to particular needs of these individuals. In this part of the questionnaire, we found some contradictions in the responses of the employees in relation to previous statements about not accepting the right to sexual life in individuals with SID and/or not creating conditions for its implementation in particular institutions. We found that most respondents inclined to implementation of sexual life according to a particular social need or desire of individuals (74.1%). According to the statements by the employees, in case that an institution allowed their individuals to practice sex life, it was mainly on the level of kissing (65.6%) and cuddling (64.5%). Another accepted form is masturbation (47.1%), which – as we mentioned before, was the most frequent form of individuals' sexual expression. Next we studied manifestations of the second level, in particular touching intimate parts of another individuals and intercourse. In cases when the employees (N = 22) did not allow implementation of sex life of their individuals, it occurred due to several reasons: there were not conditions for it (limited capacity, no premises, etc.) (5%), majority was children (1.9%), individuals do not have a need (1.5%), and there was not a representation of individuals in terms of sex (0.4%).

The attitude of employees to sexuality in individuals with SID

Whilst questioning the employees about their reactions to individuals' sexual behavior, we sub grouped responses in the questionnaire (and also in the observation coding sheet), in accordance to the levels that we had formed. Their overview is introduced in Table 17.

STAFFREACTIONS			
ELIMINATIONOF SEXUALITY	TOLERATIONOF SEXUALITY	ACCEPTANCEOF SEXUALITY	CULTIVATIONOF SEXUALITY
Punishment, Rebuke Other activities	The first level manifestations Masturbation Intercourse	The first level manifestations Masturbation Intercourse	Dialogue, Explanation

Table 17. Categorization of reactions of employees to individuals' manifestations of sexual behavior[5]

http://dx.doi.org/10.5772/55782

Response	Number	Expressedin%
Acceptance of sexuality	71	27.4
Cultivation of sexuality	64	24.7
Tolerance of sexuality	28	9.8
Elimination of sexuality (Verbal rebuke, warning)	24	9.3
Elimination of sexuality (Diverting attention – ergo therapy, game, medications)	16	6.2
I don't know	16	6.2

Table 18. Reactions of employees to individuals' manifestations of sexual behavior.

Whilst questioning the employees about what reaction to manifestations of sexual behavior of individuals with SID they considered right, we came to the following conclusions supported by statements of the employees (Table 18):

Most respondents were in favor of acceptance of sexuality in these individuals. On the second place the employees expressed tendencies towards cultivation in their opinions, which is a positive finding, even though percentage of the answers (less than a third) in terms of a necessity to educate individuals with SID in the area of sexuality, is not very encouraging. „We have discussed with a client about an intercourse, about an influence on her as well as the other individuals. We have also spoken to a partner and explained a pregnancy issue, an influence on other individuals, etc." The next in line were the reactions on the level of tolerance (ignoring – neglecting, let it take its course, no reaction), and a relatively high percentage were for elimination of sexuality – on the level of punishments (verbal), as well as elimination of sexuality by diverting attention of individuals through other activities. „Ignoring, as reproving in front of the group isn't effective, the client doesn't understand instructions when she's told not to do it, that it is inappropriate in public. When masturbation and stripping take longer, I dress the client and try to draw her attention away from masturbation through various educational activities – I give her a book, puzzle."

We perceive the presented findings negatively. For this reason, we also used the observation-coding sheet to study reactions of the staff. Next, we will compare the results obtained from the questionnaire with employees' reactions to sexual behavior of individuals in SCHs.

As shown in Table 19, most employees react negatively to manifestations of sexuality of individuals with SID especially through punishment or rebuke. Second in line are reactions on the level of tolerance of sexuality manifestations of the first level, and masturbation. Toleration of an intercourse was registered only in one respondent's statement. It is likewise with reactions on the level of acceptance, where employees are most often able to accept the first level manifestations of sexual behavior, less often they accept masturbation, not in one case do they tolerate an intercourse. Reactions on the level of cultivation of sexuality were registered only in one case. While comparing results acquired from the observation coding sheets with results presented in the questionnaire, we suggest that employees tend to evaluate their reactions more positively in general evaluation, a more negative attitude we found while dealing with a particular registered sexual expression of individuals directly in contact with them in an institution.

Code	1		2		3		Total		Nospecification	
	N	%	N	%	N	%	N	%	N	%
a	89	19.4	11	2.4	-	0.0	100	21.8	0	0.0
b	43	9.4	41	8.9	1	0.2	85	18.5	1	0.2
c	54	11.8	26	5.7	0	0.0	80	17.4	0	0.0
d	1	0.2	0	0.0	0	0.0	1	0.2	55	12.0

a1 – elimination of sexuality through punishment, a2 – elimination of sexuality through other activity, b1 – tolerance of the first level manifestations, b2 – tolerance of masturbation, b3 – tolerance of coitus, c1 – acceptance of the first level manifestations, c2 – acceptance of masturbation, c3 – acceptance of coitus, d – cultivation of sexuality

Table 19. Reactions of employees to registered sexual expression of individuals.

Opinions of employees on sexuality in individuals with SID

As it is necessary to address the problem of sexuality of individuals with SID, and as we are aware of the presence of inappropriate behavior, which could be a consequence of cumulated energy, we were interested which manifestations of individuals' behavior in SCHs have employees come across with. The most common manifestations of cumulated sexual energy according to the respondents are: quick mood swings, aggression, neurotic, and other inappropriate social behavior (head banging against the wall, hands biting). Others are stereotypical movements, neurotic drooling or sexual symbolic substitutions. There were also statements claiming that such behavior is not manifested, and other answers where the respondents stated – cannot identify (4.2%), insomnia (1.2%), the others with an occurrence of 0.8% – crying, questioning, enhanced nervousness and seeking quarrels, rubbing intimate parts against a bed or sofa, harassing other individuals, and agitation (0.4%). Our questionnaires also contained questions concerning opinions of respondents on sexuality in individuals with SID in general. Our further questions concerned opinions of employees on sexual motivation (sexual impulse) and sexual maturity in individuals with SID, with whom they are in direct contact. As our results show, most questioned regard sexual motivation in individuals with SID the same as in individuals without a disability. However, in the answers, we often found statements that sexual impulse in these individuals is reduced (27.4%) or increased (16.6%), and 25.9% could not answer this issue. We also found that most employees did not have enough information in this area. In relation to sexual maturity in individuals with SID, most respondents believe that sexual maturity occurs in people with SID later (36.3%), a relatively high percentage could not answer this question (29.7%), and some believe that it occurs earlier (17.4%). These results show, that awareness of employees about issues of sexuality in individuals with SID is at a low.

Masturbation activities and an intercourse in individuals with SID

A part of research on sexual behavior mostly relates to the frequency of masturbation. In the current study, we provide results studying masturbation activities – as detected by directly asking employees as well as data obtained through anonymous questionnaires.

Particularly positive are responses of the staff about masturbation as a normal phenomenon (57.1%) or even useful (releasing sexual tension) (52.9%). It is likely to be perceived as a sign of a positive trend and recognition of reality, that in relation to unsatisfactory conditions for partners' sexual life in SCHs (see above), individuals really have no other choice. Actual behavior or a form of sexual auto stimulation (masturbation) was specified by staff as follows: client themselves touch vagina or penis, rub a sex organ against an object, use some sex aids. In the option "other" they stated I don't know, he/she masturbates in privacy, I'm not interested, I don't have an experience.

It should be noted that masturbation is a manifestation of healthy sexual expression and it should be supported in case it is manifested in a way that does not bother the others or it is not performed in a self-harming way. In regard to a positive perception of masturbation, we were interested on employees' reaction to such manifestations in individuals with SID. We provide an example of a statement of a SCH employee:

A 32-year-old male with moderate ID.*The client was touching his private parts and smiling while doing it. He pulled his pants down and masturbated on a chair in his room. Then he laid down on the floor flat on his stomach and rubbed his genitals against the floor, and he was laughing while breathing loudly.*

Reactions of employees to masturbation are, similar to above analyzed reactions to sexual behavior in general, mostly on the level of tolerance (34.3%) (i.e. they do not solve, take no notice or let presented behavior take its course), and acceptance (33.6%) (they perceive masturbation as a natural manifestation and need that releases tension, it is allowed to individuals in privacy). We rarely found cultivation of sexuality (thus advising individuals, engaging in conversations, sex education, regulation of masturbation in public in connection with acceptance of masturbation) (around 18.1%). We evaluate negatively reactions on the level of elimination of masturbation, either through punishment or through diverting their attention to other activities – a prayer, work or relaxation (4.6%).

One of the means of a restrictive "solution" of sexuality in individuals with SID is the use of drugs that can reduce the ability to masturbate and experience libido (e.g. antipsychotics). While unethical, this method is still applied in many institutions (21.6%). Based on our experience, obtained from interviews with both professional caregivers and psychiatrists, we concluded that the situation in SCHs is even worse. We have also come to the conclusion, that most respondents similarly dismissed the idea of on allowing an intercourse (49%). Statements of professional caregivers and individuals themselves support an interest in the intercourse in individuals with SID:

A 37-year-old female with moderate ID.*The client needs surveillance; she is moody, sometimes even impulsively aggressive – once she hit another client in a conflict. Her behavior is infantile,* *problematic for insufficient regulation, a possibility of short-circuit behavior. Everyday she wants to see a penis and without educators' surveillance she would be able to carry out an intercourse. Kissing, touching intimate parts of the opposite sex and manifestations of sexual impulse in verbal communication and spontaneous stripping in front of other individuals, occur on a daily basis.*

A 40-year-old male with SID.*The client has a girlfriend and they cuddle and kiss. Another female client takes care of him, she brings him lunch, pours coffee, etc. He spends almost all his time with her, they listen to music, go for a walk holding their hands. In the past, they had an intercourse, too.*

To find more information on partnerships of individuals with SID in SCHs, we further questioned if they had a stable partnership. According to the staff in most institutions, they do not form a stable partnership. The reasons for the lack of stable partnership included, that individuals swap partners (they don't care for a stable partner) (8.9%), an age category doesn't correspond (under or over age) (6.2%), an institution doesn't create conditions (5%), they form only friendships (5%), a more SID (4.6%), non-coeducation (2.7%), and the lack of support by the institution (0.8%).

Homosexual orientation and activities in individuals with SID

There has also been a lot of research done recently focusing on the prevalence of homosexuality and homosexual experiences in the general public [25]. A few research studies have implied that the prevalence of homosexuality in individuals with a disability is similar to the general public. For example, several British studies have confirmed, that individuals with ID (mostly men) have in fact the same sexual contact with the same sex as the others, but it is hard for people in their environment to accept it [22]. Therefore, we were interested in a situation among individuals with SID and opinions of employees on these complex issues. Whilst seeking opinions of employees on a sexual interest of individuals in a person of the same sex, the findings of other authors are being documented and suggest, that even in individuals with SID an interest in a same-sex person occurs. In many responses of employees, such interest comes both from men and women (38.2%), some think it comes more from men (26.3%), and only 7.3% indicate that it comes only from women. These results are perhaps a reflection of the specific surrounding in which individuals are situated, as well as possibilities to meet up with the opposite sex (so called pseudo homosexual relations). Despite what has been presented, the results imply that it is sensible to expect that homosexual orientation or homosexual activities exist among individuals with SID. Under such circumstances in SCHs, it might be very difficult for a young person with SID to find support for his/her homosexual identity.

In relation to the results on the prevalence of interest in homosexuality among individuals with SID, it is not surprising to find that most respondents wrongly perceived homosexuality as an illness (45.2%). Next (with relatively smaller representation) is homosexuality perceived by respondents as a variation (23.6%). The other responses sound pejorative and are not consistent with up-to-date knowledge and current views of homosexuality by the society.

Reproductive – contraceptive behavior in individuals with SID in SCHs

According to all available information about sexual behavior in humans, over the last decades, a dramatic increase in a number of individuals using some forms of contraceptives at an intercourse has occurred in most countries around the world. This is related to a variety of social and cultural factors that have changed the attitudes of whole society to reproductive behavior. Planned and desired parenthood is applied more and more often. In the current study, we analyzed the most common birth control methods in women with SID and evaluated opinions of SCHs' employees in this area. Specifically, the respondents gave the following answers: none (we don't deal with it) (12%), I don't know (5.4%), they don't have an intercourse so nothing can happen (3.1%), separated bedrooms (2.7%), constant surveillance (2.7%), we prevent an intercourse with the opposite sex (2.3%). The presented methods of preventing unplanned pregnancies are perceived negatively, as it is a violation of the rights to the freedom of sexual manifestation of individuals with SID.

The optional use of contraceptives (N = 45) included: condom (6.2%), an intrauterine device (4.2%), contraceptive pills (2.7%), and sterilization (2.3%). These results reflect low prevalence of contraceptives use and an occurrence of sterilization that is currently an obsolete and with etiquette irreconcilable form of birth control. They also point to negative attitudes of employees to the questions of parenthood of individuals with SID. It confirms the results acquired within the attitudes questionnaire and also within the opinions of employees on the rights of these individuals to have children.

One of the consequences of not using contraception is undoubtedly an unplanned pregnancy. Fortunately, most employees have not encountered an unplanned pregnancy in SCHs. The employees who have encountered this problem (N = 22), suggested: abortion (68.2%) or the client's transfer to another institution (4.5%).

Opinions of employees on sexual abuse of individuals with SID

The problem of sexual abuse has gained increased public attention all over the world in the past few decades. In our research, we have focused on detecting prevalence of sexual abuse of individuals with SID in SCHs. We found that that most employees have not encountered problems in individuals in SCHs (66.8%). However, 23.5% alert to presence of sexual abuse among individuals with SID. Employees, who have encountered sexual abuse, further identified the sexual abuse offender. In most cases the offender was another client on a higher intellectual level (22%), a client on a lower intellectual level (6.2%), other relatives (4.6%), and strangers (1.2%). We also registered a SCH employee (0.8%) and even a parent of a client (0.4%) in the responses.

According to opinions of employees, the most commonly identified forms of sexual abuse were inappropriate touching of a client's body only, a client was forced to perform oral sex, an offender demanded that a client stimulate him with a hand, an offender demanded coitus in the anus, an offender demanded coitus in the vagina, others were registered with an occurrence of less than 5%.

In accordance with criminal law, which rigorously determines procedures detecting sexual abuse, we were interested how employees react to such situations. Most respondents deal with a report of sexual abuse through an interview with a client (23.6%). The respondents then indicate a possibility of reporting sexual abuse of a client to the management of an institution (12.4%). In an option other the questioned gave these statements: most individuals with SID don't communicate, they cannot express their emotions and impressions, they moved the client to another part of the building of the SCH, a report to a paramedic, the client didn't report it, stronger surveillance and preventing the situation from happening again by communication with the client and a psychiatrist.

Finally, our results show high latent criminality in the area of sexual abuse. Our results indicate that only a small number of incidences (whether it was sexual abuse or sexual harassment) were reported to the police. In most cases sexual abuse occurred in milder forms (inappropriate touching, masturbation), only a smaller number involved penetrative behavior of an offender (a vaginal or anal intercourse). Also from this point of view, occurrence of sexual abuse rising latent criminality and severity of such kind of crime in population of individuals with SID, too, is particularly alarming.

An analysis of professional training of SCHs' employees in sexuality

The other dimension, which we investigated in terms of analysis of the quality-of-life in individuals with SID in SCHs, was the degree of awareness of employees and individuals on questions of sexuality and sex education. At first, we present evaluation of awareness of individuals with SID by employees. The respondents were asked to evaluate it on a five-point scale in a questionnaire.

In the responses of the employees, we found that a negative opinion dominates on sex education and enlightenment of individuals with SID. Despite minimal awareness of individuals in the area of sex education, the majority did not find it necessary to increase awareness in this group of individuals (63.3%). The same negative conclusion bring the data which reflect, that most respondents have not completed any professional training in the area of sexuality or sex education of individuals with ID.

The respondents who completed professional training (N = 66) were asked to evaluate its benefit on a five-point scale. Responses to this question showed the quality of employees' preparedness to solve problems of sexuality in individuals with SID in SCHs. The participants considered completed training as beneficial.

The respondents were given an opportunity to evaluate their own awareness in the questions of a partnership and sexuality of individuals with SID in the following questionnaire item (again, evaluation was done using a five-point scale). We found out, that most respondents stated that it was average.

The presented conclusions of employees' education in the area sex education suggest a necessity to improve the situation by implementing continuing education ensuring the quality of training to solve the complex problem of individuals with SID. Sexual socialization of individuals with SID is in fact accomplished through symbols and patterns, submissions, expectations, rules and deputations, particularly from SCHs' employees.

An analysis of the situation in the area of sex education and its implementation in SCHs

Sex education has become a public issue around the world including the Slovak republic. Political conservatives and religious fundamentalists have often voiced their displeasure against open sex education. According to Murphy [15], the main arguments of opponents of sex education are statements that sex education is inefficient and is responsible for greater sexual engagement in adolescents or early sexual initiation. Yet, according to the majority of research, the opposite is correct.

We found that the ratio of number of the respondents' answers commenting on advantages versus disadvantages of sex education was around 2 (161 advantages: 84 disadvantages). The employees found advantages in better awareness of individuals, improvement of sexual behavior, prevention, improvement of communication with a partner, sexual self-acceptance, releasing sexual tension, etc. Disadvantages include stress, creating sexual tension, early interest in sexual intercourse, etc. The opinions of employees show differences in responses, which prove a low degree of preparedness to solve sexuality issues in individuals with SID in SCHs. Finally, according to the World Health Organization sex education leads young people in sexual activities to adopt safer sexual practices, postpone sexual initiation and decrease overall sexual activity. At the same time it declares, that the study has showed neither increase nor decrease of sexual activity in young people under the influence of sex education [25].

Other various international studies have monitored the main information population sources about sexual issues. The responses showed, that the main sources of sex education in individuals with SID were employees (57.9%) followed by the family, and educational institutions, e.g. SCH. Professional caregivers generally prefer sex education implementation at school to family.

The current status in sex education in SCHs

Here we studied the particular situation in SCHs in the implementation of sex education. Therefore, in the next questionnaire item, employees were asked whether there was an appointed person or a supervisor in the institution who was in charge of dealing with sexuality and partnerships issues. Unfortunately, in most SCHs there were no such persons. This finding indicates low (if any) degree of sex education in SCHs. Only 7.7% of the employees stated, that the institution has an appointed person (e.g. the director, appointed employee, head of nursing or social division, doctor, head nurse or psychologist).

Another finding revealed that with the exception of two institutions, there were no rules or guidelines used in SCHs, which would instruct on how to deal with basic issues occurring in an institution in the area of sexual and partnerships in individuals with SID. According to Kozáková [7], it is essential for SCHs to have exact guidelines on how to manage basic issues occurring in this area. As we mentioned before, such a practice is standard in many countries and essential for successful sex education implementation.

According to several published studies, sex education is implemented only in a small number of public schools [25]. We were interested in situations in SCHs and found most institutions do not perform sex education.

12. Conclusions

Clearly, evaluation of sexuality and partnerships in people with SID in SCHs will significantly help to improve quality of life in these individuals. Based on our research, we may state that, in the area of sexuality and sex education in individuals with SID, there are still many gaps. We believe that through understanding opinions and attitudes of employees, it would be possible to get a better picture about the problem. Therefore, we hope that the present study will serve as a tool for further discussions in the area of sexuality and sex education in individuals with SID in SCHs. We consider our conclusions as the beginning of long-term research of sexual behaviors that would enable us to precisely study the main characteristics of sexual behavior in individuals with SID, changing opinions and attitudes of employees and other specialists, who are in touch with these individuals, and last but not least, to study how the changes reflect in the quality of provided sex education and its reflection in quality of life in individuals with SID.

Finally, it should be added that manifestations of sexuality in individuals with SID would be more complicated if a person with ID lives within a system with no support for sexual knowledge and relations, and if he/she has not acquired correct and accurate information about their sexuality and sexual presentations. It is important, that employees and all others who participate in education of individuals with SID understand the specifications of the sexuality in people with SID. Only then, we can open the way for effective sex education in these individuals.

Acknowledgements

This study is a part of the framework of research task VEGA 1/0942/11 Improving the quality of life of individuals with a more severe intellectual disability in social care homes in dimensions of sexuality and partnerships.

Appendix

1. We regard individuals with more SID as individuals with moderate and SID in this paper.
2. Coding of the observed sexual manifestations:

The sexual behavior in individuals-to- employees: V1 – cuddling of same-sex employees, V2 – cuddling of opposite-sex employees, V3 – stroking of a same-sex employee, V4 – stroking of an opposite-sex employee, V7 – desire to look attractive to a same-sex employee, V8 – desire to look attractive to an opposite-sex employee, V9 – blushing in the presence of a same-sex employee, V10 – blushing in the presence of an opposite-sex employee, V11 – manifestations of joy, happiness in the presence of a same-sex employee, V12 – manifestations of joy, happiness in the presence of an opposite-sex employee, V13 – kissing of a same-sex employee, V15

– sexual assaults (attacks), V17 – masturbating in the presence of an employee, V19 – V22 other atypical sexuality manifestations

3. The p-value is the level calculated from the Mann-Whitney U-test and represents the probability of error caused by accepting the hypothesis of difference between tested variables. If $p > \alpha$, we accept the null hypothesis H_0; if $p < \alpha$, we reject the null hypothesis and accept the alternative hypothesis H_1.

4. Pearson correlation coefficient r (1) falls between [-1, 1], (2) if $|r| = 1$, then all points are lying on a straight line, (3) if $r = 0$, we call X and Y uncorrelated variables, (4) if $r<0$, then Y decreases and it indicates a negative association, (5) if $r>0$, then Y increases and it indicates a positive association.

5. *Reactions on the level of elimination* of sexuality and sex education of individuals with SID – we have as reactions, which hold an opinion that if a sexual need and desire occur in individuals, it is necessary to keep them occupied with entertainment or work, so they do not have time to think about these needs, or its fulfilling is prevented by a punishment (verbal).

Reactions on the level of tolerance – we have as reactions, which tolerate their sexual needs or desires and their satisfying through masturbation. Employees are aware of sexual needs, they tolerate them, but they do not provide any support in this area (mostly on the level of ignorance).

Reactions on the level of acceptance – employees are aware that individuals have the same sexual needs as the others; they are equally sexually active, "sexually normal".

Reactions on the level of cultivation – we have perceived as the highest level, when employees are aware that individuals not only have the same sexual needs as the others, but they also need open sex education, support and counseling in this area and their task is to provide the support.

Author details

Stanislava Listiak Mandzakova

Department of Special Pedagogy, Faculty of Education, University of Presov, Presov, Slovak Republic

References

[1] Conod, L, & Servais, L. Sexual life in subjects with intellectual disability. In: La vida sexual de las personas con discapacidad intelectual. Cuernavaca: Salud Pública de México, 2008, Cuskelly M., Gilmore L. Attitudes to sexuality ques tionnaire (Individ-

uals with an intellectual disability): Scale development and community norms. In: Journal of intellectual & developmental disability, (2000). vol. 3, no. 3, , 50(2), 214-221.

[2] Diederich, N, & Greacen, T. Enquête sur la sexualité et la prévention du SIDA chez les adultes handicapés mentaux en Ile de France. In: Revue Européenne du Handicap Mental. (1996). , 3(9), 20-32.

[3] Divišová, J. Postoje personálu pracujícího s osobami s mentálním postižením vůči sexualitě- srovnání s Australským modelem. In: Sexualita mentálně postižených- II.: sborník materiálů z druhé celostátní konference organizované o. s. ORFEUS. Praha: Centrum denních služeb o.s. ORFEUS, 2009, Divišová J. Attitudes of staff working with individuals with intellectual disabilities towards sexuality- comparison with an Australian model. In: Sexuality of individuals with intellectual disabilities- II.: Book of proceedings of the second national conference organised by ORFEUS civic association. Prague: Daily services centre ORFEUS civic association, (2009). p92-97.], 92-97.

[4] Gust, D. A, Wang, S. A, Grot, J, et al. National survey of sexual behavior and sexual behavior policies in facilities for individuals with mental retardation/developmental disabilities. In: Am J Ment Retard, (2003). , 5(6), 365-373.

[5] Kochová, H, Bernátová, R, & Palková, V. Aplikácia informačno-komunikačných technológií do edukačného procesu prírodovedy na 1. stupni základnej školy. In: Nové technologie ve vdělávání. Olomouc: Univerzita Palackého v Olomouci, (2011). Kochová H., Bernátová R., Palková V. Application of information and communication technologies into the educational process at Science lessons at the first level of primary school. In: New technologies in education. Olomouc: Palacky University in Olomouc, 2011, p15-20.], 15-20.

[6] Kozáková, Z. Výchova k bezpečnosti a sexuální výchova osob s mentálním postižením. In: Domácí násilí a zdravotně postižené: sborník materiálů z celostátní konference. Prague: HB Print, (2005). Kozáková Z. Safety and sex education in individuals with an intellectual disability. In: Domestic violence and individuals with disabilities: Book of proceedings of the national conference. Prague: HB Print, 2005.]

[7] Mandzáková, S. K niektorým otázkam kvality života klientov domovov sociálnych služieb v oblasti sexuálnych a partnerských vzťahov. In: Mandzáková S. (ed.) Kvalita života osôb s mentálnym postihnutím v domovoch sociálnych služieb: zborník z vedeckej konferencie s medzinárodnou účasťou. Prešov: PF PU v Prešove, (2010). Mandzáková S. Onsome questions about quality of life of clients in social care homes in terms of sexual and partner ships. In: Mandzáková S. (ed.) Quality of life of individuals with ID in social care homes: Book of proceedings of the international scientific conference. Presov: Faculty of Education, University of Presov in Presov, 2010, p74-81.], 74-81.

[8] Marková, D. Predmanželská sexualita v kontextoch sexuálnej diverzity a variability. Bratislava: Regent, (2007). Marková D. Premarital sexuality in terms of sexual diversity and variability. Bratislava: Regent, 2007.]

[9] Marková, D, & Truhlárová, Z. Sexualita u osôb s mentálnym postihnutím. In: Sexualita mentálně postižených II.: sborník materiálů z druhé celostátní konference organizované o. s. ORFEUS. Prague: Centrum denních služeb o.s. ORFEUS, (2009)., 45-56.

[10] [Marková, D, & Truhlárová, Z. Sexuality in individuals with an intellectual disability. In: Sexuality of individuals with intellectual disabilities- II.: Book of proceedings of the second national conference organized by ORFEUS civic association. Prague: Daily services centre ORFEUS civic association, (2009)., 45-56.

[11] Matulay, K. Mentálna retardácia. Martin: Osveta, (1986). Matulay K. Mental retardation. Martin: Osveta, 1986.]

[12] Mccabe, M P. Sex Education Programs for People With Mental Retardation. In: Mental Retardation Journal, (1993)., 131(6), 377-387.

[13] Mcgillivary, J. A. Level of knowledge and risk of contracting HIV/AIDS amongst young adults with mild/moderate intellectual disability. In: J Appl Res Int Dis, (1999)., 12(4), 113-126.

[14] Mellan, J. Biologické hlediská sexuality mentálně postižených osob: Somatopsychické základy sexuality. In: Sexualita mentálně postižených: sborník materiálů z celostátní konference. Prague: Centrum denních služeb o.s. ORFEUS, (2004). Mellan J. Biological aspects of sexuality in individuals with mental disabilities: Somatopsychic bases of sexuality. In: Sexuality of individuals with intellectual disabilities: Book of proceedings of the national conference. Prague: Daily services centre ORFEUS civic association, 2004, p10-16.], 10-16.

[15] Murphy, N. A. Sexuality of Children and Adolescents With Developmental Disabilities. In: Pediatrics, (2006)., 118(1), 398-403.

[16] Oakleyová, A. Pohlaví, gender a společnost. Prague: Portál, (2000). Oakleyová A. Sex, gender and society. Prague: Portal, 2000.]

[17] Prevendárová, J. Výchova k manželstvu a rodičovstvu. Bratislava: SPN, (2000). Prevendárová J. Marriage and parenthood education. Bratislava: SPN, 2000.]

[18] Pueschel, S. M, & Scola, P. S. Parents perception of social and sexual functions in adolescents with Down syndrome. In: J Ment Defic Res, (1988)., 32(6), 215-220.

[19] Raboch, J. Očima sexuóloga. Prague: Avicenum, (1998). Raboch J. Through a sexuologist's eyes. Prague: Avicenum, 1998.]

[20] Šedá, V. Zamyšleni nad sexualni vychovou mentalně postiženych. In: Sexualita mentálně postižených: sborník materiálů z celostátní konference. Prague: Centrum denních služeb o.s. ORFEUS, (2004). Šedá V. Thoughts on sex education in individuals with intellectual disabilities. In: Sexuality of individuals with intellectual disabilities:

Book of proceedings of the national conference. Prague: Daily services centre ORFEUS civic association, 2004, p71-74.], 71-74.

[21] Škorpíková, A. Partnerský a sexuální život osob s mentálním postižením: diploma thesis. Brno: Masarykova univerzita PF, (2007). Škorpíková A. Partner and sexual life in individuals with an intellectual disability: diploma thesis. Brno: Masaryk University- Faculty of Education, 2007.]

[22] Šurabová, I. Partnerské vzťahy a sexualita občanov s mentálnym postihnutím. In: Informácie ZP MP v SR, (2002). vol. X, Šurabová I. Partnerships and sexuality in subjects with an intellectual disability. In: Information by ZP PM in the SR [Civic association helping people with intellectual disabilities in the Slovak Republic], 2002, vol. X, no. 50-51, p4-7.](50-51), 4-7.

[23] Tóthová, M. Sexuálna výchova a príprava na partnerstvo pre ľudí so špecifickými potrebami. In: Klimek Ľ., Klohna B. (eds.). Láska, partnerstvo, erotika a sexuálny život telesne postihnutých občanov: zborník prednášok. Prešov: ZOM, (2001). Tóthová, M. Sex education and partners' preparation for people with specific needs. In: Klimek Ľ., Klohna B. (eds.). Love, partnership, eroticism and sexual life in subjects with a physical disability: Book of proceedings of lectures. Presov: ZOM, 2001, p9-20.], 9-20.

[24] Walker-hirsch, L. The Facts of Life... and More: Sexuality and Intimacy for People with Intellectual Disabilities. Baltimore: Paul. H. Brookes Publishing, (2007).

[25] Weiss, P, & Zverina, J. Sexuální chování v ČR- situace a trendy. Prague: Portál, (2001). Weiss P., Zvěřina J. 2001. Sexual behaviour in the Czech Republic- situation and trends. Prague: Portal, 2001.]